Praise for *The AI Instinct*

"Once humans and machines become a coupled system, the security problem changes. You're no longer securing software, you're securing decision-making. *The AI Instinct* explains why that matters."

—Bruce Schneier
Author of *Rewiring Democracy: How AI Will Transform Politics, Government, and Citizenship.*

"Artificial intelligence will give people access to the knowledge and talents of human intelligence across the globe. It's an extraordinary time in our history, and it requires new ways of understanding what is happening and where it all leads. In this book, AI pioneer Rana Gujrai moves beyond the stereotypical discussions and adds the missing pieces that are essential to gaining valuable insights into this unstoppable technology. He explores how AI potentially signals the next stage in human evolution as we merge with our technologies. How will this co-evolution change us? What type of society will we create? What are the pitfalls in this journey? We all share the same future. This book shows us what is achievable and what still needs to be done."

—Tom Foremski
Silicon Valley Watcher

"Offering a rare blend of technical clarity and philosophical depth, The AI Instinct redefines artificial intelligence as a force that reshapes the way we think, feel, and decide, rather than as a mere tool at our disposal. The book is about intelligence—both human and artificial—as well as emotions, perception, and learning from experience. Rana Gujral's concept of Artificial General Experience is both provocative and practical, providing a solid foundation for big ideas about AGI and human–AI hybrids by linking them to real-world cognitive science and lived experience."

—Dr. Paweł Gora
Founder and CEO, Quantum AI Foundation

"The AI Instinct makes a case the field needs to hear: that the experiential dimension of AI deserves as much serious attention as the capability dimension. Gujral's vision of superintelligence as fusion rather than replacement is both more plausible and more interesting than the standard narrative."

—Cameron Berg
Founder and Director, Reciprocal Research

THE AI INSTINCT

The Future of AI and HUMAN Decision-Making

RANA GUJRAL

Foreword by KAI-FU LEE

WILEY

Published by John Wiley & Sons, Inc., Hoboken, New Jersey.

Library of Congress Cataloging-in-Publication Data has been applied for:

Hardback ISBN: 9781394392124
ePDF ISBN: 9781394392148
ePub ISBN: 9781394392131
oBook ISBN: 9781394392155

Cover Design: Wiley
Author Photo: © Alfiya Ganieva

Typesetting: Set 10.5/13pt Palatino LT Std by Lumina Datamatics

SKY10161190_070126

For my mother. Your faith never left me.

Contents

Acknowledgments

A book is a strange thing. It feels solitary while you are writing it, but it is never truly written alone.

First, I want to thank Joscha Bach. Since the early days, his work and our conversations have shaped how I think about intelligence, experience, and the stories our minds construct about reality. This book carries his influence in ways that are obvious to me, even when they are invisible on the page.

I am also deeply grateful to Kai-Fu Lee for writing the foreword and helping set the tone for what follows.

To the team at Behavioral Signals, thank you for carrying the load with me. I am lucky to work with people who care about building things the right way.

To Wiley, thank you for making this real. Lindsey Givens, thank you for the initial push and for the coordination that kept everything moving, including guiding the cover design process. Navin Vijayakumar, thank you for leading the effort inside Wiley and for being the steady hand throughout. Tracy Brown, thank you for the line and structural edit and for helping shape the structure and flow. Colleen Diamond, thank you for the copy edit and for making the book clearer, tighter, and stronger, and for the precision and polish.

To Jessica Kantor, thank you for being a rare kind of collaborator. You helped me shape the narrative, find the clearest language, and stay true to what I actually meant. This book is stronger because of you.

To my family, thank you for your patience, your humor, and your love through the many rounds of writing and rewriting. Piggie, thank you for your quiet, steady support and love. Some of the best ideas came on long walks with my son, where we kept returning to the same topics: reality, consciousness, free will, and what any of it means.

And to the readers and community who have followed these ideas over time, asked sharp questions, and encouraged me to keep going, thank you.

—Rana Gujral

About the Author

Rana Gujral builds at the frontier of artificial intelligence (AI), cognitive AI, and the long-horizon path toward artificial general intelligence (AGI). As CEO of Behavioral Signals, he turns advances in speech-signal processing and representation learning into production systems that infer intent, emotion, and risk from voice. His work sits at the intersection of voice intelligence and high-trust decisioning, powering real-time interaction intelligence, trust and safety, and deepfake detection for enterprises operating in high-stakes environments.

Rana's technical interests span paralinguistics, self-supervised speech embeddings, transformer-based modeling, Artificial General Experience (AGE), robotics, and cybersecurity, with a consistent focus on moving ideas from research into systems that ship.

Rana speaks at global forums including the World Government Summit, World AI Cannes Festival, The Next Web LEAP, and TEDx, and has written for TechCrunch and Forbes on applied AI, voice intelligence, leadership, and responsible deployment. Recognition includes Inc.'s "AI Entrepreneur to Watch," SoundHound's "Top 40 Voice AI Influencers," "Most Influential CEO" by CEO Monthly, and "Top 10 Entrepreneurs to Follow" by Huffington Post.

Foreword

In the foreword to my own book, *AI 2041: Ten Visions for Our Future,* I invoked Alexis de Tocqueville's warning to the French Parliament that the most dangerous moments in history often arrive quietly, when societies fail to recognize that foundational shifts are already underway. At the time, I was speaking about the economic, geopolitical, and social consequences of artificial intelligence. In this book, Rana Gujral sounds a different kind of alarm, one that is no less important. His concern is not simply what AI will do to jobs, nations, or markets, but what it will do to human cognition, experience, and agency itself.

As someone who has spent decades building AI systems and studying their impact, I recognize the urgency of this question. We are entering a phase where AI no longer sits outside us as a tool, but increasingly participates in how we think, decide, and understand the world. Gujral's work challenges us to look beyond familiar debates about capability and control, and to confront a deeper issue: how intelligence, experience, and humanity may co-evolve in the years ahead.

For years, the public conversation about AI has swung between extremes. On one side is optimism bordering on hype, the promise that AI will solve nearly every problem and usher in effortless abundance. On the other side is fear, the belief that AI will replace us, deceive us, or eventually overpower us. Both narratives miss what is already happening in front of us. AI is not only something we use. It is becoming embedded in our habits, our institutions, and our patterns of judgment.

That is why this book matters now. The most consequential changes brought by AI rarely arrive as dramatic ruptures. They arrive as defaults. In business, decision pipelines increasingly surface AI recommendations before a human has fully formed an opinion. In consumer technology, attention systems learn individual psychology with unsettling precision. In healthcare, finance, education, and defense, the same pattern repeats. AI enters first as support, then as efficiency, and eventually as invisible infrastructure. When that happens, the central question is no longer whether AI is accurate. The question is whether humans remain meaningfully in control.

Gujral is not writing from the sidelines. He has spent years building and deploying AI systems that operate close to the most human layer of technology, the layer where behavior, emotion, and intent matter. If you have worked in technology long enough, you learn that the systems that change society most are not always the ones with the most impressive benchmarks. They are the ones that reshape human habits. In an industry that often treats human impact as an afterthought, Gujral treats it as a design constraint.

What distinguishes this book is that it goes beyond analysis and raises new questions about the direction of intelligence itself. It invites readers to reconsider the relationship between intelligence and experience, to look beyond the assumption that progress necessarily leads toward fully standalone machines, and to think differently about what superintelligence might actually mean. Whether one agrees with every conclusion or not, these perspectives encourage a more careful and more grounded conversation about where AI may be taking us.

There is also a broader context that makes these questions especially urgent. AI is not being developed in one place, under one set of norms, or for one shared purpose. It is being shaped across different political systems, cultural assumptions, and competitive pressures. It is a geopolitical technology as much as an economic one. Nations will pursue it for growth, security, and influence. Companies will pursue it for advantage. Individuals will pursue it for leverage. In such an environment, the most likely outcome is not catastrophe. It is drift.

A gradual outsourcing of judgment. A quiet weakening of human agency. A slow normalization of systems that shape our choices without our awareness. None of this requires coercion. Convenience is enough. Speed is enough. Optimization is enough.

I first met Rana at an event where we spoke about the trajectory of AI and what it would mean for human life beyond productivity. We stayed in touch, and later I joined Behavioral Signals as a scientific advisor. Over the years, what has consistently stood out to me is that Rana approaches these questions with a rare balance of practicality and philosophical range. He understands the real constraints of deploying AI in complex environments, with real users and real consequences. At the same time, he is willing to engage with questions many technologists avoid, questions of identity, agency, meaning, and what it will feel like to live alongside systems that increasingly resemble minds.

AI will bring extraordinary progress. It will also bring new forms of dependency. It will widen gaps if we let it, and it will strengthen institutions if we design it well. The outcome is not predetermined. It will be shaped by choices made by engineers, executives, policymakers, educators, and citizens. We need new frameworks because terms like AGI have become containers for too many meanings. When a concept means everything, it cannot guide engineering, policy, or public understanding.

As you begin this book, I invite you to read it with two questions in mind. They are not technical questions, but human ones.

First, what would it mean for AI to move from solving problems to shaping experience.

Second, if superintelligence emerges through the coupling of human and machine intelligence, or if it does not, what responsibilities do we inherit as designers, leaders, and citizens living alongside these systems.

If you keep those questions close, you will take more from this book than information. You will take a clearer way of seeing and planning for the world that is arriving, and a sharper sense of what it will require to navigate it wisely.

—Kai-Fu Lee
CEO, 01.AI; Chairman and CEO,
Sinovation Ventures; TIME100 AI (2023)
Author of *AI Superpowers* (New York Times bestseller)
and *AI 2041* (Best Book of the Year, WSJ)

What Does This Book Cover

The AI Instinct offers a comprehensive and interdisciplinary examination of the evolving relationship between human cognition and AI, with particular emphasis on the ways in which emerging forms of machine intelligence are reshaping decision-making, social organization, and the conceptual foundations of human identity. It situates contemporary developments in AI within the longer intellectual history of cognitive science, philosophy of mind, and behavioral research, while also extending the discussion toward future trajectories that include artificial general intelligence (AGI), Artificial General Experience (AGE), and the potential emergence of hybrid forms of cognition that integrate biological and computational processes.

The early chapters establish a theoretical foundation by exploring the nature of human cognition, including the historical development of cognitive science and the relevance of dual process theories that distinguish between fast, intuitive modes of reasoning and slower, more deliberative forms of thought, as well as the central role of emotion in shaping judgment, perception, and action. These human cognitive frameworks are then placed in dialogue with contemporary approaches to AI, including cognitive, behavioral, and emotion-oriented systems, in order to clarify how machine architectures both emulate and diverge from human mental processes across a range of real-world applications. Through detailed discussion of current deployments of AI in sectors such as healthcare, defense, finance, customer engagement, and security, the book demonstrates how systems already participate in environments that demand complex forms of perception, interpretation, and decision support.

Building on this empirical and conceptual grounding, the book examines the ongoing pursuit of artificial general intelligence and introduces the concept of Artificial General Experience as a complementary framework for understanding machine systems that increasingly operate across domains of perception, learning, and interaction. It analyzes how the development of more general and adaptive AI systems may influence human decision making, institutional structures, and social practices, particularly as distinctions between intuitive and deliberative forms of machine behavior become more salient in high stakes contexts. The discussion then turns toward emerging models of human–AI hybrid cognition, including collaborative decision-making systems and biotechnological integrations such as neural interfaces and brain–computer technologies, with attention to how these developments may reshape domains ranging from medicine and security to creative practice, political life, and interpersonal companionship.

The later chapters engage directly with the ethical, psychological, and societal implications of emotionally responsive and behaviorally adaptive AI systems, including concerns about influence, autonomy, identity formation, and the long-term consequences of sustained interaction with machines that simulate or instantiate affective capacities. Drawing on perspectives from sociology, cultural anthropology, and related fields, the book examines how different societies may interpret, resist, or integrate hybrid forms of intelligence, and how these responses may reflect deeper cultural narratives about agency, embodiment, and what it means to be human. The concluding sections extend this analysis beyond AGI toward the possibility of artificial superintelligence (ASI) and more radical forms of technological and biological synthesis, including the prospect of codified consciousness and nontraditional embodiments for both human and artificial agents, while offering a critical reflection on how individuals and institutions might responsibly prepare for a future in which human evolution is increasingly entangled with the trajectories of intelligent machines.

Who Should Read This Book

We are entering a historic moment in which AI no longer merely retrieves information but increasingly shapes attention, interpretation, and belief, thereby participating in the formation of individual judgment and collective sense-making.

This book is written for readers who seek conceptual clarity, theoretical language, and analytical models for understanding what is already unfolding in human–machine relations, as well as for anticipating the social, cognitive, and ethical consequences of what lies ahead. It is addressed to those who have grown dissatisfied with reductive debates that frame the future of AI as a binary contest between imminent AGI and trivial pattern matching, between total alignment and total loss of control, or between human replacement and mere technological augmentation, and who instead seek more precise and intellectually serious ways of thinking about intelligence, agency, and experience.

It is for:

Scientists, engineers, and researchers who require a grounded framework for interpreting AI beyond benchmark performance, extending analysis into domains of perception, emotion, agency, and experience that increasingly define real-world human–machine interaction.

Philosophers, cognitive scientists, and curious minds concerned with foundational questions regarding the nature of human experience, the conditions of choice, and the stability of identity in contexts where cognition becomes distributed across human and AI systems.

Practitioners and system builders working in Cognitive, Behavioral, and Emotion AI, particularly those engaged with questions of trust, intent, and influence, who seek to understand the deeper implications of machine mediated judgment for individual autonomy and institutional responsibility.

Leaders, founders, and business decision-makers operating in high-stakes environments in which AI already shapes strategic outcomes, resource allocation, and human behavior, and who require a rigorous lens through which to evaluate risk, accountability, and normative standards when human and machine judgments become increasingly entangled.

Students and younger generations who will inhabit a world defined by sustained human–AI partnerships, for whom the question is no longer whether intelligent systems will shape cognition and social life, but how conceptions of agency, responsibility, and the meaning of human experience can be articulated and defended in an age of machine mediated perception and decision making.

For readers who seek to understand not only how AI functions, but how it is reshaping what it means to think, to choose, and to be human, this book offers a coherent framework for critical engagement with the next phase of cognitive and social transformation.

How to Contact the Author

We appreciate your input and questions about this book. For updates, resources, and to contact the author, visit theaiinstinct.com. You can also learn more at ranagujral.com, and find Rana on X at @RanaGujral.

CHAPTER 1

Introduction: Our Definition of "There" Is Wrong

I'm what you would call a classic nerd. I got my first degree in computer science, have worked in tech almost my entire life, and am the CEO of an AI company. I unwind by reading big-idea books on computation, cognition, and the nature of reality. And yet, like many people who genuinely love big ideas, I also get bored reading books about highly important topics. Over time, I have learned to treat any book as a kind of choose-your-own-adventure; otherwise, I risk getting stuck halfway through and never finishing it at all.

David Deutsch's *The Fabric of Reality* (Allen Lane/The Penguin Press, 1997),[1] one of many books that shaped the thinking behind the pages you are about to read, is a perfect example. I read Chapters 1 through 9 as normal, then straight to the last chapter. After I read the second-to-last chapter, working in reverse order for a few chapters, I then went back to where I left off at the end of Chapter 9 and continued through the chapters I had not yet read. By the end, I was entirely convinced that the way I read the book was exactly how it should be experienced. I tell you this not as a recommendation but as a confession. I assume you might approach this book the same way.

If you're anything like me, you're standing in the middle of a bookstore or on a seller's page reading the introduction, wondering what you're going to get out of this book. You might even flip through to

[1] Deutsch, D. (1997). *The Fabric of Reality: The Science of Parallel Universes — and Its Implications.* Allen Lane/Penguin Press.

the final chapter, which will be entirely confusing to you if you don't go back through and read those middle chapters at some point, and may gasp at the audacity of my claims. Please do; then come back to the beginning.

Why This Book, Why Now

We are living through a quiet but consequential transition. Artificial intelligence (AI) is shifting from something we use to something we think with. It is no longer just producing outputs. It is shaping the inputs to our minds: what we read, what we see, what we believe, what we feel, and what we do next. The most important question in front of us is no longer whether a standalone machine will wake up one day and surpass us. The question is whether we will sleepwalk into outsourced judgment and engineered instinct, or whether we will deliberately build a partnership in which intelligence and experience expand together.

This book exists now because that fork exists now. The systems we are building are already influencing human behavior, decision-making, and identity. The choices we make in this window will shape not just what AI becomes but what humans become alongside it.

Three Ideas That Frame the Journey

For my readers who are looking for something new in the AI conversation, this book introduces three unique ideas:

The first is Artificial General Experience (AGE), which, without giving too much away, embarks on the understanding that intelligence without experience is hollow and that in AI, until we measure the path to intelligence and the path to experience side by side, "Artificial general intelligence (AGI)[2] when?" is the wrong question.

The second is the creation of hybrid cognition and fused minds, not the continued conversation of "us versus them" but the blend of human biology and machine intelligence, shifting from isolated minds and standalone models to shared hybrid systems that shape how humans think and live.

[2] We will dive into this term and its definition(s) in future chapters, but for now consider it to be a theoretical, advanced form of AI that is as "intelligent" as, or more so, than humans.

The third is a fundamentally different understanding of artificial superintelligence (ASI) entirely. Not just how it manifests, but who or what is actually superintelligent, and whether superintelligence manifests as an external entity at all.

Together, these ideas offer a way out of the stale dichotomies that dominate the current discourse. As our final chapter shows, the question "How do we get there?" in terms of AI might be the wrong question entirely. Perhaps our definition of "there" is wrong. We see endless debates about whether large language models (LLMs) can reach AGI, whether we need world models, entirely new learning techniques, or a complete overhaul of existing paradigms. Layered on top of this is the deeper debate about the nature of the mind itself. Is the brain an antenna or a generator? Where does cognition emerge from? What counts as the substrate of the mind, and how does that shape what is possible to build?

Some argue that you cannot replicate human consciousness or general intelligence without duplicating the biological substrate itself. Others believe that much of what we call mind can be computationally modeled, that emotion, expression, and even sentience may arise from the right architecture. Some scientists argue that none of this is as mysterious as it looks, that we can build feeling models, experience models, and perhaps one day, consciousness models. But there remains the counterview that what makes us special is not only the software of mind but the organic substrate through which it flows. The brain may not produce consciousness, but there may be something essential about it that enables consciousness to manifest. The fusion idea resolves this tension. If the substrate is special and the models are computable, then putting them together gives you something new. You can leverage the unique qualities of the organic substrate while also breaking past its limitations through computational systems. This is where superintelligence becomes plausible, not as a disembodied alien mind but as a fused mind, a hybrid agent that draws simultaneously from biology and computation.

The Missing Axis

When you look at most AI development, especially the cultural conversation around AGI, the axis we obsess over is capability. Bigger models. Better benchmarks. More reasoning tokens. Better planning.

Better tools. Better world models. The implicit assumption is that if we push competence high enough, something like general intelligence will "arrive."

But in practice, the failures that worry me most are not the dumb failures. They are the confident failures inside systems that people trust. A model can be persuasive without being right. It can rationalize without understanding. It can optimize a proxy while missing the point. It can repeat the same class of mistake in new clothing because nothing in it remembers consequence the way a living system does. You'll soon read in the text that the definition of intelligence has as many possibilities as the definition of AGI.

That is when it became clear to me that we were missing a second dimension. Not just the path to intelligence, but the path to experience. The epiphany for AGE didn't arrive as philosophy; it arrived as self-observation.

I noticed a subtle shift. I was no longer using AI as a separate tool, like a calculator or timer. Instead, I was using it the way I use my own mind: to draft thoughts, test arguments, frame decisions. Using AI this way felt like productivity, but I saw in it a deeper implication in that once a system sits inside the loop of attention and judgment, its most important property isn't how clever it makes you sound in that instance. It's what using the system does to you over time. Does it create continuity, better beliefs, or better instincts? Or does it create a convincing stream of outputs that feels like thinking without actually building judgment?

That's when the missing axis snapped into view. Humans don't just solve problems. We accumulate experience. We update from consequences. We learn patterns across time and internalize cost. If AI is going to be embedded in human decision-making, then "intelligence" isn't enough. The system needs a relationship to time, context, and consequence. This realization was the seed of AGE.

What would it mean to build systems where the primitive isn't the answer, but the experience? Systems organized around a continuous stream that binds perception, memory, intent, and consequence into something coherent, the way living minds do. Not memory as storage, but memory as transformation: the system becomes different after it is wrong, after it causes harm, after it sees a pattern play out in the real world. Not intelligence as isolated competence, but judgment as an accumulated relationship with consequences. That is the missing dimension in the AGI conversation. We keep measuring intelligence as if it is the whole story. But what matters in human contexts is whether the system can participate in continuity of experience.

For readers who simply want to understand AI, this book is also for you. This text provides a digestible understanding of AI in its entirety, from how it was created by modeling human cognition, to what cognitive AI is and how it works, to how it's currently being used in the world across different sectors, to what AGI is and how the heck we plan to get there.

In the following chapters, you will journey through not only AI's inner workings but the human mind itself, asking a deceptively simple question: before we predict where AI is going, do we even understand what it means to think? Readers discover why humans rely on two dueling modes of thought, fast intuition and slow reasoning, and how AI was created by replicating one of those systems of thought. The text brings AI down from the abstract and into the real world, showing how deeply it has already threaded itself into everyday life. It begins with the historical arc that carried us from the AI winters to the deep learning revolution and then to modern transformer-based models that can reason, plan, and act. What once required supercomputers and specialized labs now sits inside the phones of farmers, students, clinicians, and call center workers. We trace how deep learning, reinforcement learning, and autonomous agents set the foundation for a world where AI is both a collaborator and an independent problem solver.

In this book, I discuss emerging collaboration models, from centaur teams to cyborg integrations and hive-like collectives, each offering different strengths depending on context. I examine the promises and risks of these structures, especially how accountability can blur when humans rely on powerful automated partners. This book looks deeply at trust, transparency, cognitive privacy, embodied interfaces, and the ethical dilemmas that arise when machines participate in our choices, senses, and even bodies. In a world where cognition is shared, what becomes of agency and the human experience? I explore the intimate terrain of a hybrid humanity and what happens to society and life as we know it. As the boundary between brain and device becomes porous, ethical questions shift from theoretical debates to lived experience. Drawing on insights from neuroscience, philosophy, posthumanism, and transhumanism, I examine how hybrid bodies and hybrid minds challenge long-held ideas about autonomy, embodiment, and even the soul.

Yes, AI is polarizing, but it is also misunderstood. Not in an emotional, teenage "you just don't get me"[3] sort of way, but in the

---

[3] Or perhaps yes in this exact way.

same way that most of us humans don't understand how our own brains work, or how our own hormones work, or how our own emotions work. All of which would make life a large degree easier if we did understand these things that have such a heavy impact on our day-to-day existence. The same is true for AI. We fear what we do not know. We give false trust to things that appear reliable. Imagine if everyone who used it in any capacity, or refused to use it in any capacity, understood how it worked, how we're getting it to evolve to meet new needs, what its current limitations truly are, and what the actual possibilities are for its long-term existence and its role in the lives of humans.

In the end, this is the hope of what we can achieve. Not a future defined by fear or inevitability, but a future shaped by intentional design, deeper understanding, and the courage to imagine that human evolution may not be ending at all. It may simply be entering a new phase, one in which intelligence, experience, and consciousness expand through partnership rather than replacement. What will we become if we succeed? That all depends on what we do now.

CHAPTER

2

The Nature of Human Cognition and AI's Approach to Mimicking It

In order to understand where humans are going, and where artificial intelligence is going, we need to understand where we are now. We need to build on an understanding of what makes humans, well, *human*, including how our brains work, how emotion guides our thinking, and how our own brains are constantly changing in response to our experiences.

At the very basic level, the brain is an adaptive information-processing system. As such, it constantly takes in sights, smells, sounds, tastes, and touch (sensory data), processes this information through networks of neurons (around 86 billion), and creates outputs in the form of thoughts, emotions, actions, movements, and more.

Different brain regions specialize in different tasks. If you're dealing with decision-making, self-control, planning, and similar items, your prefrontal cortex handles these tasks. If you're looking at emotion, survival instincts, or memory, these tasks are handled in your limbic system.

The default mode network (DMN) is what works with introspection and self-related thought. This network is a collection of brain regions that becomes most active when we're not focused on a specific task—when we're daydreaming, reflecting on the past, imagining the future, or thinking about ourselves and others. It's the brain's "background mode," supporting things like self-awareness, moral reasoning, and our sense of identity. It's the quiet hum behind our stream of consciousness, often running when everything else seems still.

Understanding how the brain works involves many disciplines—linguistics, philosophy, anthropology, computer science, neuroscience, and, most recently, artificial intelligence (AI). In this section, we'll trace the ideas related to these disciplines, from early philosophy through modern cognitive science.

From Philosophy to Cognitive Science

Humans have always been interested in understanding how the brain works. It was Hippocrates in around 400 BCE who pushed the idea that the brain, not the heart, was the one responsible for emotion, understanding, and thinking. He believed that the brain controlled the functions of the body. In the seventeenth century, René Descartes claimed "I think, therefore I am," proposing *mind–body dualism*, which philosophizes that the mind and body are distinct, separate entities. However, it was in the twentieth century when these philosophies and theories all took substantial root and became a study of their own: cognitive science (CS).

In the 1950s, CS became a distinct field of interdisciplinary study, focusing on the mind and its processes. This was partially because the study of psychology and behaviorism rejected the idea that internal mental processes are based in science and should encourage scientific discussion. At the time, behaviorism focused primarily on observable behaviors and largely ignored internal mental states. However, as researchers continued their explorations of how humans process information, solve problems, and even use language, they realized that internal mental representations (memories, thoughts, etc.) should be included in this study. This was the *cognitive revolution* and was pioneered by scientists like Noam Chomsky,[1] Herbert Simon,[2] and Jerome Bruner.[3] These scientists and others confirmed the idea that the mind could be systematically modeled and measured.

The beauty of cognitive science is that it brings together insights from many disciplines: linguistics, philosophy, anthropology, computer

[1] American professor and public intellectual known for his work in linguistics, political activism, and social criticism.

[2] American scholar whose work influenced the fields of computer science, economics, and cognitive psychology.

[3] American psychologist who made significant contributions to human cognitive psychology and cognitive learning theory in educational psychology.

science, neuroscience, and, most recently, AI. CS allows scientists to view the mind not just through a set of behaviors, but as an information-processing system. Further, with advances in technology and imaging, scientists can observe the brain in action and give empirical weight to cognitive models. Scientists now understand that cognition is not linear; it is shaped by our environments, our physical bodies, and our interactions with others.

Systems of Thought: Kahneman and Beyond

Daniel Kahneman's *dual-process thinking* is a core concept within the field of CS. We'll be diving much deeper into this further along in this chapter, but in cognitive terms, dual-process thinking says the brain is constantly switching between two modes, depending on context, energy levels, and time pressure. The first mode, System 1, governs most of our daily actions and is fast, automatic, and intuitive. It operates automatically and quickly, with little or no effort and no sense of voluntary control. This system is relied upon heavily but can sometimes be sloppy in its efforts. Walking, recognizing people, and even doing basic math all come from System 1, which often involves operating below the level of conscious awareness. The second mode, System 2, is used when solving problems or reflecting deeply, and is slower, effortful, and logical. Analyzing an argument, doing complex math, or even parallel parking your car are all done in System 2 and involve focusing your attention on specific details. System 2 allocates attention to mental activities, such as complex computations, that demand more effort. This mode takes more effort but is often more accurate than System 1, which *often* gets to the right answer faster, but not always.

The *computational theory of cognition* is central to modern CS and proposes that the mind operates like a computer. It advances the idea that cognition can be understood as rule based, symbolic, and computational. While the computational perspective is not universally accepted, scholars like Joscha Bach[4] support this theory and argue that cognition is likely computable. If cognition truly is computational and governed by symbolic rules, then in principle, it should be replicable in artificial systems.

---

[4] Cognitive scientist, AI researcher, and philosopher of mind who has worked extensively on modeling cognition using computational systems.

Conversely, if cognition were rooted in something mystical or metaphysical, existing in an inaccessible fourth or n-dimension,[5] it would likely defy replication. However, current scientific evidence increasingly supports the view that cognition operates within structured, algorithmic frameworks. Elan Barenholtz[6] invites a rethinking of cognition as software. He proposes a general theory of cognition grounded in autoregression. He puts forth the idea that every thought, every image, every decision is just the brain's next best guess and that there is no fixed self—just a rolling loop of internal outputs, continuously feeding forward into themselves.[7] His works align closely with Bach's framework around consciousness as a computational architecture.

As small proof of this, CS has evolved in tandem with advancements in AI, machine learning, neurotechnology, and behavioral AI. This is because artificial systems were created based on how the human brain works, computationally, and as they grow they become more adept at mimicking human behavior.

If cognition is fundamentally computational, guided by symbolic processing, algorithms, and structured information flow, then this reinforces the idea that it can, at least in part, be modeled and recreated in AI. However, that does not mean we have access to the kind of substrate capable of supporting its full replication. Just as early computers lacked the hardware to simulate even basic neural activity, we may not yet have the materials, energy systems, or dimensional understanding needed to recreate the full complexity of human cognition. This does not suggest that cognition is beyond science. Rather, it may exist within a computational paradigm that remains out of reach due to current limitations in technology or understanding. Scholars like Bach acknowledge that replicating the full phenomenological experience may require unknown levels of abstraction, architecture, or complexity that we are not yet capable of constructing. While we are not there yet in terms of tapping into these substrates, I am

[5] Dimensions beyond our familiar three spatial dimensions (length, width, and height). In physics, the fourth dimension is often considered time. In mathematics, the fourth dimension is an additional spatial dimension, orthogonal to the other three, enabling further freedom of movement within that space. N-dimensional space is a broader mathematical concept that generalizes the idea of dimensions to any number (n).

[6] American researcher using behavioral and embedded computational approaches to study the brain and behavior with the goal of developing a broad theoretical framework of neural function.

[7] We'll discuss this further in Chapter 5 "Human-AI Hybrid Cognition: The Future of Decision-making."

inclined to believe that it's possible in the future. CS now bridges the gap between natural intelligence and synthetic intelligence, offering critical insight into what it means to be human.

Emotion, Bias, and the Limits of Rational Models

Something that is very human, and many are likely thankful not yet prominently found in interactions with AI, is emotion. In the past,[8] experts and philosophers believed emotion to be a contaminant to rationality and a source of error or noise in decision-making. *Decision theory*, a framework for making choices particularly under uncertainty, came from disciplines like economics, philosophy, and mathematics and originally put forth the notion that humans are rational agents, calculating and making decisions without emotion, intuition, and psychological variables. Over time, it became clear that this was not an accurate depiction of how the human mind works nor how the vast majority of humans make decisions.

In 1955, Simon published "A Behavioral Model of Rational Choice," which introduced the idea of bounded rationality and how humans are not perfectly rational but instead make decisions based on information, cognitive capacity, and time constraints. He proposed that individuals don't always strive for optimal results but rather seek solutions and outcomes that are deemed satisfactory. Kahneman and Amos Tversky[9] took this even further in the 1970s when they published "Judgment Under Uncertainty: Heuristics and Biases," and ultimately created Prospect Theory, which dismantled the rational agent theory by showing that humans very clearly and often deviate from logical decision-making due to biases and heuristics, mental shortcuts or "rules" that people use to simplify complex decisions and judgments. While the layman may think that bias is a bug in the human operating system, it's actually a survival optimization.

[8] Some may still believe this, and you can still often see it in misogynistic political messaging that paints women as poor leaders because they're deemed as "too emotional" which is inaccurate on both accounts.

[9] Israeli cognitive and mathematical psychologist and a key figure in the discovery of systematic human cognitive bias and handling of risk.

Bias, Heuristics, and the Limits of Rational Choice

However, it wasn't until the 1990s when Antonio Damasio,[10] through the Somatic Marker Hypothesis, broke down the idea that emotion is a threat to rationality. He proposed that it is a central component of how individuals evaluate risk, make judgments, and ultimately make decisions. Damasio researched brain-injured patients who had damaged emotional processing centers. His work showed that people couldn't make even the most basic of decisions without emotion and ultimately reframed emotion as an essential input, not a flaw in the decision-making system.

Further, in 2015 Jennifer Lerner[11] published "Emotion and Decision Making"[12] which showed that emotions are not distractions or noise, and instead act as "potent, pervasive, predictable" drivers of judgment and choice and are capable of both facilitating decision-making and impairing it. The research dove deeply into the idea of task-specific emotions or integral emotions, which include feelings that arise in direct connection to the situation; this could be fear or anxiety when evaluating danger or pride after completing a task. The work argued that these emotional responses are meaningful data points within the human decision-making system and help guide, instead of disrupting, ultimately calibrating both split-second judgments and deeper, more reflective decisions over time.

The Somatic and Emotional Foundations of Decision-making

A key mechanism in the research are appraisal tendencies, specific emotions, and emotional states that trigger patterns in how we process information and decision-making. Things like anxiety naturally will cause the average person to sharpen their sensitivity to threats. Anger increases certainty and generally causes people to make riskier decisions. Sadness increases attention to detail but can also lead to more conservative choices. These inclinations are not from random impulses

[10] Portuguese neuroscientist whose research in neuroscience has shown that emotions play a central role in social cognition and decision-making.

[11] American experimental social psychologist known for her research in emotion and decision theory.

[12] Lerner, Li, Valdesolo, and Kassam. (2015). Emotion and Decision Making. *Annual Review of Psychology* 66:799–823. https://doi.org/10.1146/annurev-psych-010213-115043.

but instead come from identifiable cognitive biases, which show that emotion has structure and influence in human decision-making.

Empirical studies have shown that how emotion guides behavior is also impacted by things like time pressures, cognitive load or overload, and mood. Risk perception shifts as emotional states become default decision inputs, so people rely more on emotion when rushed or when they're feeling mentally strained, they're more likely to make quick, emotionally guided decisions.

Embodied Cognition and Distributed Intelligence

Through recent work in somatic neuroscience, epigenetics, and cellular memory, there is some scientific evidence that suggests memory is not only stored in the brain but throughout the body. This means that the body itself "remembers" experiences through things like hormone regulation, gene expression, and nervous system sensitivity. Embodied memory implies that imprints of past experiences can be passed down across generations without awareness.

Developmental biologist Mike Levin[13] has taken this further, showing that information can be stored and processed at the cellular and even tissue-level through bioelectric signaling. His research suggests that cells use electrical gradients to "decide" how to grow, repair, and organize themselves, almost like a form of problem-solving. An example from his research is that of planarian flatworms regenerating heads with altered body patterns, and proving that this process is influenced by previously stored bioelectrical states, not solely by their DNA. This implies that a kind of memory or intelligence exists in the body's tissues, independent of the brain. If memory and decision-making can happen at the level of cells or tissues, it reframes the idea of cognition itself. Intelligence may not be a top-down process originating in the brain, but a distributed phenomenon that arises from many layers of biological computation.

Going further, embodied cognition is a theory that advances that thought and cognition are deeply influenced by the body's physical interactions with the world, rather than being solely a function of the brain. The theory suggests that physical experiences, including our

---

[13] American developmental and synthetic biologist who works at the intersection of developmental biology, artificial life, bioengineering, synthetic morphology, and cognitive science.

sensations and movements, shape our cognitive processes, including how we understand and interact with the world, and even how we form abstract concepts. Thought extends throughout the entire body, encompassing the nervous system, gut, immune responses, hormonal feedback loops, and even the external environment. When people talk about having a "gut feeling" or sensing danger without clear evidence, they're often experiencing the embodied cognition system at work. This is done through your gut, nervous system, and hormonal networks which are constantly processing vast amounts of information, most subconsciously. These systems can detect subtle changes in the environment and trigger physiological responses before your conscious mind catches up.

Damasio's somatic marker hypothesis also bridged the connection between feeling and thinking, and suggested that emotions are not just mental states but also involve physiological responses in the body, and he proposed that consciousness emerges from the dynamic interaction between the brain, body, and environment.

These decentralized views of memory and intelligence expand the frontier of what might one day be replicated or augmented in artificial systems. Current AI models lack embodiment, which might limit current intuitive contextualizations. This also raises a profound question: If the human body can think in its own way, do we need to rethink what it means for something to be considered intelligent?

A powerful example found in humans is that of studies of Holocaust survivors and their descendants. The children and grandchildren show measurable physiological and psychological effects like higher cortisol, heightened anxiety, and increased vulnerability to PTSD, despite not having lived through the experience themselves. These impacts don't come from stories or learned behavior, they're found in epigenetic modifications, where traumatic stress alters the way genes are expressed. This has implications for our understanding of cognitive bias. If certain emotional or behavioral responses are shaped by inherited physiological patterns, then bias may, in some cases, precede conscious thought. This doesn't mean that bias is hardwired in a fatalistic sense, but that we may carry preloaded emotional filters shaped by the survival strategies of our ancestors.

Neuroplasticity and the Malleable Brain

Still, in most cases, bias is shaped by environment and experience. We're not born with political views, sports team loyalties, or cultural preferences; we acquire them from our families, peers, and social environments. As we grow older and gain independence, we start exerting more control over our biases. We choose what to focus on, what information to trust, and which viewpoints to adopt or reject.

Neuroplasticity is the brain's remarkable ability to reorganize itself and form new neural pathways and connections throughout one's life. While development of the brain is known to occur between birth and early to mid-twenties, it is never done changing. It remains malleable and responsive to everything from injury to our environment to our habits, hobbies, behaviors, emotions, and even thoughts. Through recent discoveries in neuroscience from researchers like Michael Merzenich[14] and Eric Kandel,[15] we now have scientific evidence that our brains change in response to our experiences.

An incredible example of this is that of substance use and addiction. While widely considered a chronic and genetic disease, research shows that the repeated use of a substance or activity develops and reinforces neural circuits around that addictive behavior, whether it be a substance addiction or process addiction.[16] These circuits are repeatedly strengthened through cycles of craving, use, and reward. The more that pathway is used, the stronger it becomes. Addiction is not just a disease or behavioral issue; it is a neural loop. A positive example of this is found in how we learn new languages or learn how to play musical instruments.

Just as those neural pathways are formed due to the repeated cycles of craving, use, and reward, the brain can be trained to create new pathways. Cognitive Behavioral Therapy (CBT) is one treatment for addiction and other behavioral health issues that help people to challenge their distorted thinking and recognize their triggers (when they may start to get a craving), practice new behaviors (stop the use), and instead use a healthier coping strategy (associating the reward with a different action or thought process). This, over time, encourages the brain to create and strengthen new neural links and pathways, ones that replace the unhealthy and impulsive reward-seeking behaviors with conscious decision-making. Eventually, the dopaminergic reward system responds entirely differently.

Mindfulness and meditation are other treatments used to address addiction and other issues, working by increasing activity in the prefrontal cortex and reducing automatic and emotionally-reactive

---

[14] American neuroscientist whose research on brain plasticity demonstrated that neuroplasticity remains through adulthood.

[15] Austrian-born American medical doctor who specialized in psychiatry and received the 2000 Nobel Prize in Physiology or Medicine for his research on the physiological basis of memory storage in neurons.

[16] Examples are gambling, shopping, love, sex, etc.

behavior, which are often found in the amygdala and limbic systems. When individuals observe urges without acting on them, the brain creates new neural pathways that support regulation and reflection over impulse.

There are many more therapies and examples, but the core of it is that repeatedly engaging in new and healthy responses, thoughts, and, most importantly, behaviors strengthens new neural pathways and offers the brain a way to unlearn dependency on a substance or behavior. This is why the type of addiction treatment is not always the main driver of success, but instead the length of treatment. The longer you participate in treatment, the more time your brain has to create new neural pathways.

This also means that our brains, whether we want them to or not, are constantly impacted and influenced by not only our behaviors but our surroundings, technologies, and the systems (1 and 2) we use. In the modern world and our current attention economy[17] we have a constant barrage of rapid interactions, which are tailored to provoke rapid, emotionally charged reactions. From 24/7 news cycles, to scrolling through feeds, to swiping between apps, and instantly retrieving information from the internet and ChatGPT. The platforms most people use daily are driven by metrics like engagement and cultivate habitual micro-interruptions that short-circuit patience, skepticism, and deep reasoning. This over time conditions the brain to favor fast, automatic responses over thoughtful analysis.

Unless we are consistently also utilizing our System 2 thinking with deliberate, effortful processing like deep reading, reflection, and complex problem solving, we are strengthening the intuitive and reflexive circuits associated with System 1 thinking and ultimately making quick reactions our default mode. Studies[18,19,20] show that an imbalance in using System 1 over System 2 contributes to a noticeable decline in attention span and the brain's capacity for sustained focus, suggesting that the digital environment is not just influencing what we think about but how we think altogether. As humans become

[17] The current system where human attention is treated as a scarce, valuable resource, and economic activity is driven by the competition to capture and maintain it.

[18] Korte, M. (2020). The impact of the digital revolution on human brain and behavior: where do we stand? *Dialogues in Clinical Neuroscience* 22(2): pp. 101–111. https://doi.org/10.31887/DCNS.2020.22.2/mkorte.

[19] Small, G. W. et al. (2020). Brain health consequences of digital technology use, *Dialogues in Clinical Neuroscience* 22(2): pp. 179–187. https://doi.org/10.31887/DCNS.2020.22.2/gsmall.

[20] Hasan, Md. Kamrul MS, MPH. (2024). Digital multitasking and hyperactivity: unveiling the hidden costs to brain health. *Annals of Medicine & Surgery* 86(11): pp. 6371–6373. https://doi.org/10.1097/MS9.0000000000002576.

more adept at scanning and skimming through information, we are losing our capacity for concentration, contemplation, and reflection.

Dependence on AI language models and recommendation systems further complicates this picture. These tools don't just assist with thought, we are starting to see that they shape it, subtly influencing what we see, what we consider, and even how we feel about it. As humans outsource more of the mental labor to machines, there's a risk that we're expanding the boundaries of the mind while losing touch with its core capacities. AI might grant us near-omniscient access to facts, predictions, and polished prose, but it may also erode the curiosity and friction that spark deep understanding. We begin to consume thought rather than engage in it.

Neurodivergence, Psychedelics, and Altered States

It's important that we reckon with the reality that human cognition is a moving target. We are starting to see an increase in individuals who are considered to be *neurodivergent*, which refers to variations in how brains develop and function which lead to different behaviors, ways of learning, and even ways of thinking. Autism, ADHD, and dyslexia are all forms of neurodivergence. While there is no specific "cause" of neurodivergence and it has been seen throughout history, there has been an increase in conversations around and understanding of it in recent years, which has ultimately helped to end the stigma associated with it. Autism is increasingly understood not as a deficit but as a different cognitive operating system, one marked by strengths in pattern recognition, memory, and logic, alongside differences in emotional and sensory processing.

Psychedelics are another item that have always been around in some form, but modern analysis and conversations have caused it to be more frequently used and accepted. One of the most surprising discoveries in psychedelic neuroscience is that compounds like DMT and psilocybin do not heighten brain activity; they dampen it. Specifically, they reduce activity in the brain's DMN, a region associated with self-referential thought, ego, and mental noise. Psychedelics appear to loosen top-down predictive models in the brain.[21] This leads to

[21] Carhart-Harris, R.L. et al. (2019). REBUS and the Anarchic Brain: Toward a Unified Model of the Brain Action of Psychedelics. *Pharmacological Reviews*. 71(3): pp. 316–344. https://doi.org/10.1124/pr.118.017160.

increased bottom-up sensory flow, which can create new pattern detection and insight or, in some cases, delusions.

What experts have learned is that by turning down the brain's usual chatter, the mind becomes more open, connected, and vividly aware. In many cases, it appears the brain is not the gateway to consciousness but the gatekeeper. The brain is often in the way. This suggests a radical possibility: the brain, in its everyday state, may be more of a filter than a window. It blocks, suppresses, and simplifies in order to function, but that very filtering may obscure deeper layers of reality and self. In this view, consciousness is not entirely generated by the brain but shaped, constrained, even edited by it. Silence, in this context, is not the absence of thought; it is the absence of interference. It is what allows previously hidden patterns to surface, what makes room for nonlinear, emotionally resonant insight.[22] The DMN's deactivation may dissolve the ego's narrative, allowing a more direct form of consciousness or perception. When the filter drops, what comes through may be less about knowledge and more about presence.

Insights often reported to be gleaned from psychedelic experiences, dissolution of ego, and unity with the universe are now understood not to be merely the byproducts of neurochemical disruption. Rather, they echo longstanding philosophical and ontological traditions that have questioned the nature of self, reality, and perception for millennia. Think of Buddha, Plato, and even Descartes. Further, Aldous Huxley,[23] in his 1954 work *The Doors of Perception*, described the brain as a "reducing valve," a filter that narrows the vast stream of sensory and experiential information to what is essential for survival. In his view, psychedelics loosen this valve and allow a fuller range of perception to flow into consciousness. Modern neuroscience has begun to echo this idea, showing that psychedelics reduce activity in the brain's default mode network, potentially granting access to patterns and sensations normally blocked from awareness.

I recently watched a fascinating documentary on Netflix called *The Lost Children*. It shares the true story of the search for four Indigenous siblings from the Huitoto tribe after a plane crash in the Colombian Amazon rainforest, a deadly place where even well-trained adults would have trouble surviving. The documentary follows two rescue efforts that were separate in the beginning and over time combined

[22] We will delve further into this in Chapter 6, "The Ethics and Societal Implications of Hybrid Thinking."

[23] English writer and philosopher who produced a prolific body of work across multiple genres.

to strengthen the chance they found the children. One group was the Colombian military, which used helicopters, acoustic loudspeakers, systematic grid-based sweeps, and military training. The second was a group of Indigenous volunteers who relied on ecological wisdom and knowledge of the rainforest. The teams looked for 39 days with no luck and no traces of the children when finally, part of the Indigenous group, one elder and the father of the children, took ayahuasca, a natural psychedelic called *yagé* by the local communities. The father experienced no visions, but the elder hallucinated for a few hours, vomited, and then told the men, "We'll find the children today." The team, having had no other leads for 39 days of searching, followed his report and went deeper into the rainforest. It was here that they discovered a trail of used diapers, scattered clothing, and partially eaten fruit and ultimately found the children, all alive. This is an astounding account, and while some may be skeptical, it truly shows the power of how ayahuasca opened the mind of the individual to ultimately receive information that previously had been blocked out.

All this underscores how fluid, embodied, and emotion-driven human cognition really is. With that in mind, we can now look at how AI seeks to parallel—and where it falls short of—these capacities.

From Logic to Intuition: How AI Mirrors Human Thought

Unlike humans, AI systems use only System 2 processing because they are designed to operate through deliberate computation, structured reasoning, and logic. System 2 encompasses all of this. In humans, System 2 is the part of the mind that carefully weighs evidence, evaluates information, and makes deliberate choices. AI models, everything from rule-based systems to deep learning machines, are built with the intent of processing large amounts of data systematically with goals around accuracy and optimization. They follow structured and rule-governed processes around identifying patterns, applying learned associations, and computing probabilities. This makes AI models a strong analog to System 2.

This could be confusing because System 1 is associated with quicker response times, and we know that AI processes data thousands to millions of data points per second compared to the human brain, which processes data in seconds and minutes. However, AI does not leap to conclusions based on instinct or emotion; it computes outputs based on algorithmic reasoning, and it just does so at a much faster speed compared to the human brain.

An interesting fact around this is that AI's use of System 2 thinking didn't happen naturally or by chance. Humans created AI and thus developed the processes that enable it: we created a machine that reflects how humans originally conceptualized intelligence. We taught AI to think like our idea of a rational thinker. As was discussed earlier in the chapter, early AI development was influenced by classical cognitive science. Researchers believed that in order to build intelligent machines, they needed to replicate rational human reasoning. Rational human reasoning is found in System 2; thus, AI was built to use System 2. This also allowed researchers to build something that allowed them to more easily define and design parameters and trust that the AI would stay within them. Nobody wanted a Skynet situation.

Fast and Slow Thinking: Humans vs. AI

A good example of a human and AI using different processing systems to accomplish the same action is recognizing someone. A person glances at someone across the room and instantly recognizes their friend, not because they analyze facial symmetry or pixel patterns, but because the brain matches the face to a deeply encoded memory in milliseconds. It's fast, emotional, and automatic. An AI system like a facial recognition algorithm approaches this task through structured computation. It processes the image using deep learning models that analyze abstract patterns across thousands of images, comparing features like eye placement or bone structure against a database. It does this with incredible speed, but the process itself is not intuitive.

Another example, which now takes place regularly thanks to Waymo, is driving through an intersection without hitting a pedestrian. An experienced driver is heading to turn right at an intersection and sees from the corner of their eye that a pedestrian is walking towards the curb to cross and intuitively slows down, anticipating they might step into the street, even if they don't have a walking sign to do so. The decision is made without conscious deliberation; it's shaped by pattern recognition, emotion, and lived experience. Differently, an autonomous vehicle (AV) uses lidar, radar, and cameras to map its surroundings, calculates the pedestrian's speed and trajectory, compares the input to pre-programmed scenarios, and determines whether to slow down, stop or continue at the normal speed. It's data driven and logical, relying on complex sensor fusion and modeling. AVs don't "understand" context in a human way. They're pattern-matching from training data or rules, not "reasoning" about intent the way a human might.

The beauty of humans is that we switch between System 1 and 2 based on the task at hand. An example of both humans and AI using System 2 to solve a problem can be found in reviewing a legal document. A person reads through a contract, identifies clauses, cross-checks obligations, and flags potential legal risks. This requires sustained attention, logical reasoning, and understanding of legal precedent, a slow, focused cognitive task. An AI scans the document, identifies entities and clauses using natural language processing, compares them to known templates or legal risk databases, and highlights anomalies. This mirrors System 2 logic, though it scales the task massively.

A second example is an interesting one: picking out flowers. For a human, this could be found in System 1 or 2, depending on the person and what the flowers are for. It can be System 1 if someone walks into a florist shop, sees a bouquet, and immediately chooses a flower based on instinct, aesthetic pleasure, or emotional resonance. There's no weighing of alternatives, no reflection, just a fast, intuitive reaction. That's classic System 1 behavior. It can be System 2 if someone is carefully considering the theme of a party, color palettes, seasonality of flowers, guest allergies, vase size, pricing, and vendor reliability. System 2 can include comparing options across different florists, thinking about symbolism, or how long each flower will last. If pros and cons are being weighed deliberately, that's System 2, thoughtful, analytical, effortful decision-making. AI would use System 2 processing to choose flowers by analyzing structured inputs such as event type, color scheme, budget, availability, and longevity. It would then apply rules and optimization algorithms to compare options and select the most suitable arrangement based on logical criteria. The decision is deliberate, data driven, and free of intuition or emotion.

When AI Learns to Feel: Emotion and Bias in Machines

As AI evolves and the researchers working with it aim to test limitations and push boundaries, they're starting to explore AI that mimics System 1 characteristics: emotionally aware systems, social responsiveness, and heuristic-based adaptations. However, these models are still built on System 2 applications because intuition and emotion are much harder to hardwire or replicate than calculation. This is a funny thought, at least to me, that AI are starting to use emotional and bias-based thinking, but they're deliberately planning it.

A human example of this could be someone with high intelligence coupled with antisocial traits or high-functioning sociopathy. Someone who does not feel emotions in the same way as others. Someone who knows they need to simulate empathy or emotional cues, without actually feeling them. Over time they likely learn, through observation and reasoning, how to smile at the right time, express concern, or mirror someone's distress, not because it comes naturally (System 1), but because they know it's socially effective or expected (System 2). You likely wouldn't know that someone is doing this when meeting them, so would it matter if you know that AI is doing it?

I'm reminded of a famous quote by Marvin Minsky,[24] "The question is not whether intelligent machines can have any emotions, but whether machines can be intelligent without any emotions." He highlights that emotions aren't separate from thinking, but rather a crucial aspect of it. We know that emotions are needed for different ways of thinking and problem-solving and are essential for humans to function intelligently in complex situations. Emotions guide our decision-making and actions in the real world. If we build a highly capable AI that lacks emotional awareness, we risk creating something akin to a human psychopath: brilliant, but dangerously devoid of empathy.

As someone who works with AI often,[25] I know that the future lies in building AI that uses System 1 processing to be intuitive, emotion-aware, and socially fluent systems. It's about making AI more effective at human-centered tasks. It's catching it up to intuition. Humans aren't logical creatures. We're post-rationalizers, and our brilliance lies in how we switch modes dynamically. The next frontier in AI development isn't just about improving logic, reasoning, or raw computation, which AI already does extremely well. It's about creating systems that can approximate what humans do effortlessly: make fast, context-sensitive, emotionally intelligent judgments, what we call intuition.

[24] American cognitive and computer scientist concerned largely with research in artificial intelligence.

[25] Over the last two decades at various companies, including Behavioral Signals (https://behavioralsignals.com) where I currently serve as CEO, I have led applied AI work, including building and deploying cognitive and behavioral AI systems across enterprise and public sector deployments.

Just as the brain uses patterns to shape perception, neural networks can learn universal patterns in emotion, across languages and contexts, to simulate nuanced human behavior. Cognitive AI,[26] Emotion AI, and Behavioral AI are the first attempts to do this and embed System 1 dynamics into machines, and they're continuously evolving.

Emotion AI, known as *affective computing*, enables machines to recognize, interpret, process, and even simulate human emotions. It uses machine learning, natural language processing, and computer vision in order to analyze things like human facial expressions, voice tones, and physiological signals like heart rate, brain activity (EEG), and skin conductance (GSR). Why? To create more intuitive, empathetic, and user-friendly systems and interactions with humans.

Because Emotion AI is designed to detect, interpret, and respond to human emotional cues, something previously thought to be uniquely human, you might think that the machine can feel. What's actually happening is that it is simulating rapid emotional inference, similar to how a person intuitively senses another's mood without needing to stop and analyze. Using System 1 and assessing tone of voice, facial expressions, or other responses, Emotion AI detects spontaneous emotional reactions, things like frustration, stress, or even happiness that happen quickly in interactions. This can be helpful, say, in a customer service call where the AI assistant notes your frustration and changes its tone or offers to help, much like a human might respond (on a good day at their job, lest we forget that human customer service agents can sometimes get frustrated as well). That's not the end of it, Emotion AI still also uses System 2 in scenarios where more careful, analytical thought is required. For example, using System 2, AI can evaluate a person's emotions over time, creating a complex model of their emotional responses.

AI can analyze a customer service agent's performance and suggest ways to improve their responses, based on a deep analysis of emotional cues, in order to improve future interactions.

By integrating both System 1 and 2 processing, Emotion AI can mimic human cognitive behavior in both quick and slow modes.

---

[26] Cognitive AI can be seen as a broad category that includes Emotion AI and Behavioral AI as components, though definitions vary across fields. While academic definitions of cognitive AI may be narrower, in commercial applications, integrating emotion, behavior, and cognitive insights offers a more comprehensive way to understand and respond to human interaction. For our purposes here, we will reference Cognitive AI, Emotion AI, and Behavioral AI as separate with the understanding that there is a large amount of overlap.

Behavioral AI: Understanding Human Patterns

Behavioral AI works to understand, predict, and respond to human behavior by analyzing behavior patterns. Using machine learning and natural language processing, it tracks things people do and infers motivations, preferences, and emotional states, instead of analyzing just static data. It takes that information on how humans interacted in past experiences and learns shortcuts to make fast decisions. Why? It allows for more personalized experiences in applications, more adaptive systems, and improved decision-making.

Similarly to Emotion AI, Behavioral AI uses and blends both systems in order to not just think like humans but behave like us as well. It mimics System 1 when it is capturing fast, automatic, and unconscious human behaviors. Things like click patterns, scrolling behavior, purchases, and even hesitations. It uses these reactive and emotionally driven behaviors to infer mood, intent, and even psychological traits. It can use these nonconscious behavioral signatures to aid in credit card fraud detection: is a purchase being made at an odd hour for the person? Is the location normal for the person? Are you interacting with the app in your usual rhythm or pattern? Behavioral AI using System 2 enters into higher-level modeling, tracking long-term behavioral trends. It might analyze what nudges or interventions influence outcomes or how a user's choice shifts over days or weeks. It will use this information to create personalization over time and update its models to reflect any changes in personality, motivation, or intent. It can use this information to help a student learn math by analyzing what types of problems they struggle with, how performance changes under time pressure, which explanations caused improvement, etc. and uses that information to optimize learning by suggesting different content or slowing down instruction in weaker areas.

The Future of Cognitive Flexibility in AI

The previous examples show that the future lies in AI architectures that fluidly coordinate fast heuristics with slow logic, mimicking the human dance, our ability to seamlessly and rhythmically shift between intuitive and analytical modes. Cognitive flexibility is a defining trait of human intelligence, and the challenge that we will face is building meta-cognitive AI that has the ability to shift modes as easily as humans do. Kahneman's concept of "thinking about thinking," or *metacognition*,[27] underscores the human ability to step back and evaluate our own mental processes. This reflective capacity allows us to regulate the interplay between intuitive and analytical thought, the very dance that defines human cognition.

As AI begins to mimic intuitive, emotion-sensitive decision-making, we must ask: if AI is developing fast-mode cognition, should we be investing in strengthening our own slower, more analytical faculties? The future of AI may help correct this imbalance, or deepen it, depending on how we choose to design, use, and relate to these systems.

While emotions might have begun as survival mechanisms, their role in our evolution has far outgrown utility. Beyond threat detection or social bonding, emotions diversify consciousness.

Emotions turn existence into something felt, not just lived. They are not merely evolutionary adaptations but aesthetic and existential tools, giving dimension, tone, and texture to our experience. Neuroscience suggests we rarely choose how we feel. Emotions often arise before conscious awareness, shaped by layers of conditioning, memory, trauma, and biology. Perhaps their deeper function is to make life vivid, to color the canvas of reality with meaning. As we design artificial intelligences, we face a critical question. Should we allow AI to develop its own emotional grammar, its own way of feeling the world, or will we impose a sanitized, human-centric model? While AI is already augmenting decisions for humans, there will be a time when AI will make decisions on behalf of a human or replace us in decision-making, while using human thinking and human emotion.[28] Are we ready for it?

---

[27] The term was coined by John Flavell, an American developmental psychologist specializing in children's cognitive development, in the 1970s.
[28] We will explore all of this further in Chapter 4, "The Emergence of Artificial General Intelligence (AGI)."

CHAPTER 3

AI in the Real World: Applications and Use Cases

When I was a student, I had to go to the library or leaf through our household encyclopedia to learn about, say, jaguars. I, myself, had to read through everything, decide which information was pertinent, and either write out everything with a pen or type my notes with a typewriter. Eventually, I was able to replace the typewriter with our family computer, which required dialing in to connect to the internet through a landline.

The internet at that time did not have the vast amount of information that it does now, as people were slowly creating web pages. Now, almost every Western student has their own laptop, and even students in rural and developing countries have smartphones, their very own minicomputers in their hands. To access information, they have access to trillions of web pages with information, and they simply need to head to ChatGPT and use a free account to ask AI anything and get as much information on the jaguar as possible in a second or less. AI will supply them with information and links to sources and explain why that information is relevant to the report they're writing and even offer to write it for them. I suspect more students accept this offer than not.

Just a decade ago, a farmer in Africa needed to understand or ask for mentoring from others in his area to learn which new seeds to plant if their crop was no longer faring well. They likely needed to go through several cycles of trial and error, which could jeopardize their livelihoods if they didn't yield positive results. Now, using AI apps on their smartphones, farmers can access information on seeds that take into account their region's soil type, seasonal changes, upcoming weather patterns, and more. AI can even help them find where to purchase those seeds for the best price and connect them to a distributor.

AI Goes Mainstream: How We Got Here

We've reached the point where AI is being used regularly by the common man. Gone are the days of AI being used exclusively by high-value technology companies or grant-funded scientists. Whether you're a student in the West or a farmer in the Global South, the majority of individuals are using some type of AI application to help them with tasks.

We've found ourselves here in large part because of the Dartmouth Conference in 1956. This summer workshop is widely considered to be a seminal event for AI as a research field, initiating an era of symbolic reasoning and computational experimentation, and sparking significant interest and funding. After several years, early enthusiasm waned as the technology hit barriers while AI researchers and the media had made inflated predictions about AI's capabilities, leading to unrealistic expectations. This resulted in the "AI Winters[1]" of the 1970s and 1980s, when funding from governments and the private sector decreased, the public grew skeptical, and research pivoted to more practical and attainable goals, like machine learning algorithms, natural language processing, and computer vision.

It was due to this pivot that a resurgence emerged in the late 1990s and early 2000s with machine learning approaches, most notably Deep Blue's[2] 1997 chess victory over Garry Kasparov, who was then world champion in chess. To achieve this victory, Deep Blue searched through millions of possible chess positions per second and used programmed logic to evaluate them. The machine looked up to 20 moves ahead, calculating many possible move sequences and their outcomes. Unlike deep learning, the system Deep Blue used was a deterministic system with fixed programming and static evaluation rules. The victory reinforced a belief that machines win by brute force. Few expected that data-driven learning (DDL) would soon eclipse hand-coded rules.

[1] The term was in reference to a "nuclear winter" and coined in 1984.

[2] A customized IBM RS/6000 SP supercomputer for chess-playing and the first computer to win a match.

The Deep Learning Revolution

AI momentum accelerated significantly when convolutional neural networks (CNNs) marked a turning point, sparking a deep learning revolution. The ImageNet dataset and the ImageNet Large Scale Visual Recognition Challenge, led by Fei-Fei Li[3] and colleagues, supplied the large, well labeled benchmark that made rapid progress measurable at scale. In 2012, AlexNet[4] competed on this benchmark and achieved a top five error rate of 15.3%, far ahead of the next best entry at 26.2%. AlexNet worked by mimicking the hierarchical processing of the human visual system, learning from raw pixels to recognize complex objects through layers of abstraction. This performance leap stunned AI researchers and funders and brought deep learning into the mainstream.

The rise of generative adversarial networks (GANs) in 2014 by Ian Goodfellow[5] and colleagues enabled highly realistic synthetic content, profoundly influencing creative domains. Consisting of two neural networks, a generator and a discriminator that compete with one another, the generator works to create data that is indistinguishable from real data. Meanwhile, the discriminator tries to detect whether the data is real or fake. Through this, GANs learn to create highly realistic synthetic content. This had a profound impact on creative industries, movies, music, and art and enabled the rise of deepfake videos and AI-generated art.

Reinforcement learning (RL), a machine learning paradigm where an agent learns to make decisions by interacting with an environment to maximize a cumulative reward, achieved a landmark milestone in 2016 when DeepMind's AlphaGo defeated Go world champion player Lee Sedol. Go, an ancient and complex board game, had previously been a challenge for AI due to the importance of intuition and strategic depth required to win. Earlier systems, like Deep Blue, relied on brute-force calculations when competing in games, but AlphaGo combined deep neural networks with reinforcement learning and developed strategies not programmed by humans. The victory demonstrated AI capabilities in learning, adapting, and planning in dynamic environments.

[3] Chinese-American computer scientist known for her pioneering work in artificial intelligence, particularly in computer vision.
[4] Developed by Alex Krizhevsky, along with Ilya Sutskever and Geoffrey Hinton.
[5] American computer scientist and AI expert known for his work in deep learning and artificial neural networks.

Then the big one: the introduction of transformer architectures in 2017 with the publication of the paper "Attention Is All You Need" by Ashish Vaswani[6] et al. reshaped natural language processing, largely supplanting earlier sequential models like Recurrent Neural Networks (RNNs) and Long Short-Term Memories (LSTMs) with a structure that relied on self-attention mechanisms. This breakthrough improved both accuracy and speed of language models and gave rise to advanced generative models such as GPT series,[7] fueling an explosion of AI systems capable of nuanced, human-like interaction and agentic autonomy. GPT models demonstrated unprecedented capabilities in producing context-aware, coherent, and human-like language, and as the systems became more capable, they laid the groundwork for *agentic AI*, systems that not only understand and generate language but also reason, plan, and act autonomously.

Autonomous Agents and Long Thinking

We're now at a very exciting place. Modern AI systems are built on sophisticated architectures that integrate large language models (LLMs) with planning modules, memory, and a suite of integrated tools. These autonomous agents can break down complex problems into manageable tasks, execute them, and discover ways to optimize their own performance without human guidance. For instance, DeepSeek recently demonstrated its ability to double its own speed through recursive self-improvement—a clear sign that we're entering a new era of self-optimizing machines.

At the core of this revolution is a shift from "fast, shallow" responses to deep, deliberate thinking, often called *long thinking*. Instead of instantly churning out an answer, modern models like OpenAI's o1 series take time to analyze and reason through complex challenges using a chain-of-thought approach. This method improves accuracy and opens the door to entirely new insights.

Autonomous agents,[8] AI systems capable of making decisions and acting independently to achieve goals, are engineered to interact seamlessly with real-world systems and are being used in almost every industry. They can pull data from the web, execute Python

[6] Indian computer scientist.

[7] Released by OpenAI between 2018 and 2024.

[8] Gujral, R. (2025). The Rise of the Autonomous Agents and the Emergence of Powerful Reasoners. LinkedIn. https://www.linkedin.com/pulse/rise-autonomous-agents-emergence-powerful-reasoners-rana-gujral-is43f.

code, and manage databases autonomously. For example, Salesforce's Agentforce platform now handles tasks like prequalifying leads and scheduling meetings through dynamic, multi-step operations, well beyond simple scripted responses.

One of the most exciting advancements is the ability of these agents to improve themselves. OpenAI's recent launch of the "deep research" feature highlights this concept. Using end-to-end reinforcement learning on complex browsing and reasoning tasks, DeepSeek discovered a method to double its processing speed. This self-driven optimization echoes the intelligence explosion concept outlined by Leopold Aschenbrenner in his paper, "Situational Awareness: The Decade Ahead,"[9] in which systems that reach a critical threshold can rapidly iterate on their own capabilities.

Traditional models were excellent at rapid responses but often stumbled on multi-step challenges. New reasoning models, like OpenAI's o1, take a more deliberate approach. They pause, break problems down into logical steps, and process each component carefully. This chain-of-thought method mirrors human reasoning, greatly improving performance on advanced tasks such as complex mathematics, coding challenges, and detailed scientific analyses.

AI Across Industries

AI, in all its forms, is being used across dozens of sectors. From finance to agriculture to education to entertainment to healthcare to energy. The list spans well over 20 sectors and AI continues expanding into niche and emerging fields every day. With the emergence of open-source frameworks, powerful tools, models, and research are freely available. Open-source has lowered the barrier to entry, enabled innovation, and created a shared foundation upon which anyone can build models. In general, I would love to highlight every modern use of AI in every sector, but for the sake of what I'm covering in this book, I'll stick to how Emotion AI, Behavioral AI, and Cognitive AI are being used in various industries, focusing on the early adopters: healthcare, financial services, customer service, and defense.

These domains are not only early adopters but also serve as critical environments where machine intelligence interacts with fundamental aspects of human experience. They involve perception, emotion,

[9] Aschenbrenner, L. (2024). Situational Awareness: The Decade Ahead. https://situational-awareness.ai.

decision-making, and intent—essential components in the study of consciousness. By analyzing how AI systems are applied in these contexts, we begin to explore not just functional capabilities but also the deeper computational analogs of self-awareness, subjective experience, and agency. These explorations bring us closer to engaging with the hard problem of consciousness, using real-world implementations as the starting point for understanding how artificial systems might reflect or replicate the structures underlying human awareness.

Healthcare

I am astounded every day to see the wide variety of AI applications used in healthcare. Whether used in mental health support, tailored interventions, or early detection, the convergence of AI and healthcare marks a significant transformation in the sector. Possessing remarkable capabilities, AI efficiently handles extensive datasets while facilitating the examination of complex patterns and relationships. AI is revolutionizing healthcare by providing insights and solutions that were previously beyond reach of conventional methods. These transformative tools offer advanced detection approaches, tailored treatments and therapies, and virtual care. They increase patients' access to treatment, lower costs, reduce stigma, and improve outcomes.

It's important to note the difference between the three areas we're discussing and where they're used in care. Emotion AI tools sense how patients feel, Behavioral AI tools track what patients do and adapt accordingly, and Cognitive AI tools handle thinking and decision-making tasks, freeing clinicians to focus on care.

Emotion AI

Emotion AI is probably the most fascinating use of AI in healthcare, especially in its use in mental health and behavioral health. Playing an increasingly critical role in suicide prevention, Emotion AI is being used to detect risk and enable timely intervention through analyzing emotional and behavioral signals during crisis hotline phone calls, in therapy sessions, and even online. Analyzing vocal tone, facial expressions, and physiological signals like heart rate, it can infer a patient's emotional state and assess subtle shifts, detecting signs of anxiety or depression that are not verbally addressed. For example, a hotline caller shows rising vocal tension and longer pauses. The system flags elevated risk and prompts the counselor to use a calming script and bring a supervisor on the line. The escalation happens in under 30 seconds. The tool doesn't replace the interactions with the therapist or crisis hotline worker but instead enhances early detection and helps them to respond with greater empathy and precision.

Emotion AI is also being used in regular therapy sessions when a patient's life may not be at risk but where the added support can be useful, such as in supporting therapists by providing feedback on patient sentiment trends over time, helping guide therapeutic approaches and how they're being received, and identifying when patients are disengaging. Using these tools over time builds additional trust between the patient and therapist, furthering reception of the treatment. This has been used for patients with anxiety, depression, PTSD, and other needs. In behavioral health care, Emotion AI can help monitor cravings, emotional volatility, and relapse risk by analyzing facial expressions, voice intonation, speech patterns, and biometric signals. Companies like Headspace and Ellipses Health provide real-time support and triage for mental health and behavioral health treatment by alerting human therapists or delivering interventions immediately, such as guided coping strategies or a mindfulness prompt.

In autism treatment, Emotion AI helps decode emotional cues for therapists, their neurodivergent patients, and caretakers, improving emotional recognition and regulation. Tools are used to perceive, interpret, and respond to the signals that neurotypical humans often misread or overlook. Autism spectrum conditions are increasingly understood not as deficits but as distinct cognitive architectures, marked by strengths in logic, pattern recognition, and memory, alongside differences in social and emotional processing. Interestingly, AI systems, especially those rooted in structured data processing and logic-based inference, mirror some of these same patterns.

With the surge of mental health disorders throughout the world since the pandemic,[10] and a continuously frustrating insurance industry in the United States, there has been an unprecedented demand on healthcare systems that have increased inadequacies of traditional models of mental health care. Disparities between supply and demand have increased the need for innovative solutions, which have been found in AI. Overall, Emotion AI enhances both diagnostic precision and therapeutic timing, allowing patients to be better heard and

[10] World Health Organization (2022). COVID-19 pandemic triggers 25% increase in prevalence of anxiety and depression worldwide. https://www.who.int/news/item/02-03-2022-covid-19-pandemic-triggers-25-increase-in-prevalence-of-anxiety-and-depression-worldwide.

understood, and revealing patterns that inform stronger and more adaptive care plans.

Managing chronic diseases like diabetes, arthritis, cardiovascular disease, and cancer often entails managing the pain, stress, anxiety, depression, and lack of motivation that come along with both their symptoms and their treatment. Emotion AI is being used to monitor patients' emotional states, detect early signs of mental health deterioration, and improve patient engagement, all of which are crucial in long-term chronic disease care and management. Tools alert clinicians when patients are overwhelmed, not following treatment plans, feeling more pain than they're vocalizing, or at risk of a mental health crisis. This is incredibly important, especially in cases like diabetes management, when emotional fluctuations have a direct impact on blood sugar or medication adherence, or in the case of a patient being elderly, nonverbal, or a child. For example, a patient with diabetes might show reduced check-ins and more negative affect during telehealth calls. The care team would receive a nudge to simplify dosing and add a short coaching session. Adherence improves over the next month.

While all these tools are in effect monitoring patients to help providers, Emotion AI is being used to help monitor patients who are in surgery, intensive care, or coma. Tools can't analyze traditional communication in these circumstances, but they can use Emotion AI to pick up on emotion-related physiological signals like microexpressions, heart rate variability, skin conductance, muscle tension, pupil dilation, and biometric changes. This can help an anesthesiologist adjust medication levels on a patient in surgery, detect early cases of delirium in post-operative patients, and even help doctors treating patients with disorders of consciousness (DoC) identify signals that show varying levels of consciousness to better inform diagnosis and planning.

Behavioral AI

Behavioral AI is being used in ways similar to Emotion AI but differs in that it models dynamic human patterns, things like how patients speak, act, move, and engage with care systems. Behavioral AI can detect subtle hesitations in speech or changes in physical movement that suggest early cognitive decline, emotional distress, or hesitancy of a treatment plan. It can flag, based on analysis of prior behavior, when a patient is likely to miss a medication dose or when they might disengage from a treatment plan, before it happens. In mental health, when patients often downplay symptoms, Behavioral AI can detect early markers of mania or depression. Overall, this is helpful because it can support both in-patient and virtual care teams by providing real-time insights to clinicians, allowing them to make informed decisions without relying on self-reports from patients.

Cedars-Sinai Connect, in partnership with K Health, is one virtual care platform that uses Behavioral AI to automate patient intake, including helping with symptoms assessment and triage. The chatbot interacts with patients to gather information on their symptoms, concerns, and needs, analyzes responses and behavior on the app, and provides information and feedback to treatment providers while updating ongoing care pathways. OmadaSpark is a Behavioral AI tool that collects data directly from the user and offers an AI nutrition agent that personalizes nutrition recommendations, most notably in managing chronic conditions. In many cases, Behavioral AI tools increase both patient outcomes and lower disengagement rates.[11,12,13]

Cognitive AI

Cognitive AI is being used in healthcare to replicate aspects of human cognition such as reasoning, learning, and problem-solving, but at massive speed and scale. This is being used across the board for administrative tasks, treatment plans, early disease detection, diagnosis, triage, and more. Two of the most fascinating and impactful ways it is being used are in oncology and pharmacy.

IBM Watson for Oncology is one example of Cognitive AI being used in identifying cancer treatment plans, although ultimately it was divested in 2022 after a variety of clinical shortcomings caused criticisms. During its runtime, it analyzed the patient's clinical history, including response to previous treatment, symptoms, current health, and more, and added in information from millions of medical texts, including clinical trial data, scientific literature, treatment

---

[11] Sadeh-Sharvit, S., Camp, T.D., Horton, S.E., et al. (2023). Effects of an Artificial Intelligence Platform for Behavioral Interventions on Depression and Anxiety Symptoms: Randomized Clinical Trial. *Journal of Medical Internet Research*. https://doi.org/10.2196/46781.

[12] Habicht, J., Dina, L.M., McFadyen, J., et al. (2025). Generative AI-Enabled Therapy Support Tool for Improved Clinical Outcomes and Patient Engagement in Group Therapy: Real-World Observational Study. *Journal of Medical Internet Research*. https://doi.org/10.2196/60435.

[13] Ghose, Guo, Li, and Dang. (2022). Empowering patients using smart mobile health platforms: Evidence from a randomized field experiment. *Management Information Systems Quarterly* 46(1): pp. 151–192. https://doi.org/10.25300/MISQ/2022/16201.

guidelines, and drug databases that the hospital or doctor might not have been aware of. It provides a ranked list of treatment options for that unique patient, each accompanied by a confidence score and supporting literature.

The reviewing physicians were able to see the reasoning behind each recommendation, such as the clinical trials or a specific journal article, that support a given treatment plan. This transparency helps doctors assess the tool's suggestions and integrate them with their clinical judgment. Because this was not used as a replacement of expertise, but a way to augment doctor's expertise in order to make the best decision for the patient in a timelier manner, doctors were able to identify shortcomings in the AI's suggestions.

Since then, other applications have been created to aid in both cancer diagnosis and treatment plans, including LYNA from Google Health AI, Tempus, and Path AI. These applications are especially helpful as many times with cancer treatment you have a small window to make a decision that could impact the life of the patient, and each patient is unique not only in how their body may respond to treatment, but their past health history and current situations that could put them more likely to have negative side effects from a specific course of treatment. Pain, delirium, lymphedema ... these are all potential side effects of treatment that many don't speak about but are very real and can impact treatment plans negatively if not taken into consideration.

In drug discovery and development, Cognitive AI uses natural language processing (NLP) to scan and pull together insights from biomedical data[14] and builds a knowledge graph for researchers and doctors to use. Humans are capable of doing this on their own, but it would take a lot more time. The AI also can analyze molecular pathways, gene expression patterns, and protein interactions, ultimately identifying biological targets[15] that may be involved with the disease that needs to be treated. It can detect subtle correlations in complex datasets that researchers might miss and can prioritize targets that are promising for intervention.

Cognitive AI can help in building a strong clinical trial group by helping select ideal participants based on biomarker profiles, predicting drop-out risks and treatment outcomes, and monitoring real-time patient data. Drug repurposing, using an FDA-approved medication that has an expired patent and is now less expensive[16] for a new

[14] Patient records, clinical trial reports, scientific journals.
[15] Things like enzymes and receptors.
[16] This is an ethical issue that I would love to get deeper into, but we will overlook for now.

purpose, is another area where it can be helpful. By analyzing and identifying shared molecular pathways between different diseases, overlapping gene expression signatures, side effects or off-target effects that might indicate additional therapeutic potential, and other items, it can more accurately identify therapeutic interventions for diseases and faster.

In 2020 during the first wave of COVID-19 when the world was scrambling to understand the virus and create treatment plans, BenevolentAI used Cognitive AI to mine biomedical literature, molecular data, and drug–target interactions. In just a few days, it identified the drug baricitinib, originally approved and used for rheumatoid arthritis, as a potential treatment for COVID-19. After it was flagged, researchers went through the proper clinical trials needed to make sure the information that the AI uncovered was accurate and safe, and it received regulatory approval several months later to treat hospitalized COVID-19 patients.

Senior Care AI

One more interesting area that is using these AI tools is in senior care, which is increasingly important as in the U.S., approximately 70% of individuals turning 65 will need some form of long-term care with ever-growing needs for specialized dementia care. Paro, an advanced interactive robot taking the form of a furry seal, was created to be used in nursing homes and hospitals. The robot responds to touch, eye contact, and sound to elicit calming and positive emotional responses in dementia patients. It aims to achieve the documented benefits of animal therapy in environments such as hospitals and extended care facilities where live animals present treatment or logistical difficulties. It has shown reductions in agitation and improvements in mood, similar to animal-assisted therapy, but without issues like hygiene or allergies.

Ryan, the companionbot, is a lifelike conversational robot equipped with multimodal emotion recognition and was created to help patients with dementia and depression. In a pilot study,[17] elderly residents with moderate dementia had continuous access to Ryan for 4–6 weeks and formed meaningful emotional rapport, and the emotionally intelligent version of Ryan was consistently rated as more

[17] Abdollahi, H., Mollahosseini, A., Lane, J.T., and Mahoor, M.H. (2017). A Pilot Study on Using an Intelligent Life-like Robot as a Companion for Elderly Individuals with Dementia and Depression. *IEEE-RAS International Conference on Humanoid Robots*. https://doi.org/10.1109/HUMANOIDS.2017.8246925.

engaging than scripted variants. These examples show that in situations that require specialized care that includes interaction, companionship, and emotional engagement, Emotion AI and Behavioral AI systems are able to reduce feelings of loneliness, stimulate cognitive and social activity, and provide routines, which ultimately help reduce workloads for caregivers and potentially reduce costs.

Financial Services and Customer Service

Financial institutions and customer service providers are surprising yet highly relevant spaces for the use of Emotion and Behavioral AI. When implementing, they're looking for improved efficiency in delivering services to their customers as well as improving the likelihood of a desired result. What are those desired results? Increased sales for call centers, better outcomes for unhappy customers, and, for financial institutions specifically, increased rates of loan repayment.

These outcomes are achieved by using voice and behavior analysis, taken from the current call and previous interactions, that allow for the creation of behavioral profiles for both the worker and the caller. Machine learning techniques allow for incoming calls to be routed to the worker that best matches the profile on the incoming caller, one whose skills and overall behavioral traits the customer will mesh well with. Additionally, Emotion AI can be used in real time to understand the emotions of the caller so that the worker can adjust their interaction or escalate to a higher level if needed.

The AI also provides dynamic coaching, whereby the platform whispers prompts to live workers while they're on a call and coaches them for the best outcome. The whispered prompts may recommend they slow down and speak more clearly, show more empathy, or guide them on how to de-escalate emotionally charged situations. This real-time behavioral support boosts emotional intelligence in the moment.

An example of this success comes from Behavioral Signals, the company where I serve as CEO. We worked with a European bank[18] that, due to the impact of the 2009 financial crisis, was taking measures to restructure non-performing loans[19] under a government securitization

[18] Behavioral Signals (2020). Personalized Agent Intelligent Routing (PAIR): Redefining bank's revenues through its call center. Behavioral Signals. https://behavioralsignals.com/personalized-agent-intelligent-routing-case-study.

[19] A bank loan that is subject to late repayment or is unlikely to be repaid by the borrower in full.

initiative. In general, the bank implemented the patented Emotion and Behavioral AI technology during its outreach to non-performing loan debtors to help with negotiation offers in restructuring the debt according to the debtor's ability to pay the loan. Post-implementation, the bank saw a 20.1% increase on active debt restructuring applications and saw a decrease in calls needed by 7.6%. Overall, we achieved $7.5M of additional restructured debt for the bank over the course of the evaluation and over $300M in total restructured debt. The bank also calculated that due to the increased success of interactions and fewer calls needed, they would be saving $1.5M per worker per year.

Palantir is another company using Emotion and Behavioral AI tools with financial institutions. One of the most interesting ways Palantir is applying these tools is in its partnership with Fannie Mae[20] and a Crime Detection Unit that combats mortgage fraud. The tool analyzes various datasets, such as mortgage paperwork, public records, and even social profiles, and applies behavioral pattern recognition in order to flag anomalies like mismatched addresses, usage patterns, and inconsistencies in applications. Classically, work like this could take analysts months, if they even recognize these fraudulent behaviors, whereas the AI tool takes seconds. Fannie Mae is implementing this technology across over $4.3T in assets.

Defense and Security

For use in defense and security, Emotion AI provides real-time affective insights for security screening and authentication, analyzing things like micro-expressions, body posture, speech tone, and physiological signals to detect stress, fear, or nervous behavior. Behavioral AI detects threats in networks and behavior, excelling at anomaly detection and threat monitoring by learning normal patterns and flagging deviations. Cognitive AI infuses strategic reasoning and higher-order decision-making into security systems, supporting decision-making, mission planning, and psychological operations. While it may seem like very basic stuff, the way AI is being used in the sector is actually very fascinating and can be unnerving. The majority of cases show that governments are using these models to augment and support

[20] Fannie Mae. (2025). Fannie Mae launches AI fraud detection technology partnership with Palantir. Press release (28 May). https://www.fanniemae.com/newsroom/fannie-mae-news/fannie-mae-launches-ai-fraud-detection-technology-partnership-palantir.

humans in making real-time decisions on threat assessment and subsequent actions.

In cybersecurity, penetration testing and threat hunting both use Behavioral AI in their efforts. Advanced AI tools use behavioral analysis to simulate attacker strategies and adapt to network defenses. It also helps detect anomalous user or system behavior that can signal a security breach, analyzing things like unusual login times, unexpected data transfers, and lateral movements across networks.

L3Harris introduced Cognitive AI, AMORPHOUS (Autonomous Multi-domain Operations Resiliency Platform for Heterogeneous Unmanned Swarms),[21] a pioneering command-and-control software platform now in active prototype development under contracts with the U.S. Army and the Defense Innovation Unit. AMORPHOUS enables operators to control multiple autonomous platforms (MAP), like drones, maritime vessels, and ground systems, across domains through a single user console. The platform supports decentralized decision-making, which enables individual uncrewed assets to perform their own tasks autonomously and make real-time tactical decisions inside the network. These MAP will be able to coordinate, negotiate tasks, and adapt their behavior without direct human input. It exhibits traits like multi-agent decision-making, situational awareness, and real-time strategy adjustment.

The company is also using Behavioral AI in a partnership with Intelligence Advanced Research Projects Activity (IARPA) for the Office of the Director of National Intelligence in the United States. HAYSTAC (Hidden Activity Signal and Trajectory Anomaly Characterization)[22] includes advanced simulations and modeling tools to characterize typical human mobility patterns, from daily routines to social network movements, and detect anomalous behaviors through signals like GPS, Bluetooth, and sensor data. This system is designed to generate automated alerts when movements deviate from modeled norms, flagging suspicious activity without manual oversight. Although it does not interpret emotions or intentions directly, HAYSTAC's ability to detect unusual patterns in behavior provides situational insight

[21] L3Harris Technologies (2025). L3Harris launches new technology to control autonomous swarms. Press release (10 February). https://www.businesswire.com/news/home/20250210036899/en/L3Harris-Launches-New-Technology-to-Control-Autonomous-Swarms.

[22] L3Harris Technologies (2023). L3Harris receives contract to advance technology for intelligence community. Press release (22 May). https://www.l3harris.com/newsroom/press-release/2023/05/l3harris-receives-contract-advance-technology-intelligence-community.

and enhances operational responsiveness in intelligence and crisis contexts.

These insights could be used to detect illicit weapons smuggling, human trafficking, or sanction evasion. Most commercial vessels follow well-known trade routes and schedules, HAYSTAC would be able to identify ships that detour briefly into unmonitored waters, turn off their transponder (AIS data), or rendezvous with another vessel in a location that doesn't match its declared route. It could also be used to track insider threats within military bases (as espionage is a very real problem for governments[23]), detect irregular troop or vehicle movements in adversary territory, or identify covert online coordination between threat actors.

IMPRINT Human Performance Modeling Tool,[24,25] used by the U.S. Department of Defense in training but not actively deployed in combat situations, is a Behavioral AI simulation-based workload modeling tool which forecasts human performance and mission success under real-world stressors like thermal stress (hot and cold), equipment malfunctions, time constraints, mobility constraints, and others. It does not, however, simulate emotional trauma or include emotional modeling. So while it may have the capacity to simulate things like an active battlefield or an earthquake, that is not what it is used for.

The Bohemia Interactive Simulations' VBS Control Editor is Behavioral AI that is actively deployed for defense use through the Norwegian Defence Research Establishment (FFI) where they use it to build a library of behavior models (behavior trees) for simulating infantry battle drills and mechanized squad tactics like fire-team maneuvers, squad reactions, and vehicle coordination. In general, VBS and VBS Control are actively deployed by over 60 countries, including NATO and NATO-friendly nations,[26] and are routinely used for training, mission rehearsal, and system evaluation.

---

[23] Melman, Y. (2025). Why are so many Israeli Jews spying for Iran? *The Economist* (14 February). https://www.economist.com/1843/2025/02/14/why-are-so-many-israeli-jews-spying-for-iran.

[24] Slimp, S., and Miller, M.E. (2024). Human Performance Modeling Architecture with HC-130J Mission application. Proceedings of the Human Factors and Ergonomics Society Annual Meeting 68(1): pp. 1555–1562. https://journals.sagepub.com/doi/pdf/10.1177/10711813241261280.

[25] Buck-Gengler, C.J., Raymond, W.D., Healy, A.F., and Bourne, L.E. (2012). Modeling cognitive tasks in IMPRINT. In: *Training Cognition* (Psychology Press).

[26] Norwegian Defence Research Establishment (FFI) (2010). Simulation-based training and behaviour modelling using VBS control. *Proceedings of the 15th International Command and Control Research and Technology Symposium (ICCRTS).*

The DARPA Insight program, also known as Manhunt Master Controller, was developed as an advanced research initiative and never fully deployed. DARPA is an advanced surveillance and tracking technology that uses Behavioral AI to locate and identify individuals in complex environments. It integrates multiple sensors, such as radar, satellite imagery, social media, and human reports, to detect high-value targets by modeling behavioral patterns and predicting likely actions. In theory, and possibly in active deployment, it could be used on battle fields to identify allies or enemies for extraction, exfiltration, or even neutralization. DARPA's Sea Hunter is similarly named but uses Cognitive AI instead of Behavioral for maritime autonomy, including anti-submarine tracking and logistical operations. It operates under sparse supervisory control; humans supervise remotely but don't control every action in real time, and it's equipped with advanced sensors, radar, and autonomy software.

Anduril's Lattice OS[27] is Cognitive AI that serves as a central AI-driven command and control network for military operations and integrates data from drones, sensors, and radars into a single integration layer where AI, machine learning, and sensor/data processing techniques are used to filter high-value information to users. Using cognitive reasoning, Lattice supports real-time decision-making, mission orchestration, and strategic autonomy at the tactical edge. Lattice connects the autonomous air systems Bolt, Barracuda, Fury, Ghost, Roadrunner, and Altius,[28] which enables a variety of intelligence, surveillance and reconnaissance mission profiles. Lattice is used for border control, military command and control, search and rescue, and securing critical infrastructure.

Ukraine is actively deploying precision targeting systems that incorporate Behavioral and Cognitive AI elements used in live conflict. In operation Spider's Web on 1 June 2025,[29] Ukraine used AI-enabled first-person view (FPV) drones programmed with target-recognition models that were trained on Russian bomber types. The drones autonomously identified and struck aircraft on runway zones, even during a communication loss. Additionally, the company Auterion is

[27] Anduril Industries (2023). Anduril's Lattice: a trusted dual use — commercial and military — platform for public safety, security, and defense. Press release: (30 July). https://www.anduril.com/article/anduril-s-lattice-a-trusted-dual-use-commercial-and-military-platform-for-public-safety-security.

[28] Frias, L. (2025). See the Anduril drones that are taking AI-driven warfare to new heights. *Business Insider* (3 May). https://www.businessinsider.com/anduril-drones.

[29] Law, H. (2025). Ukraine just demonstrated What AGI war could look like. *TIME Magazine* (5 June). https://time.com/7291455/ukraine-demonstrated-agi-war.

actively supplying Ukraine with AI-guided drone strike kits that enable autonomous target tracking during the final approach phase, enhancing accuracy and resisting jamming even when signal control is lost, allowing them to autonomously track and hit targets that are up to 0.62 miles away.[30] U.S. facial-recognition company Clearview AI has also provided its tools to Ukrainian officials to use in identifying Russians on their soil as well as Ukrainian collaborators.[31] AI facial recognition has been used to locate abducted Ukrainian children, identify and prosecute members of pro-Russia militias, and process citizens who lost their identification. Although this software is AI, it does not use Emotion, Behavioral, or Cognitive AI. I find Ukraine's use of the technology to be fascinating, however, so I wanted to include it here.

Companies like Taigusys and start-ups such as EmoKit Tech have developed Emotion AI-powered systems capable of inferring emotional states from facial micro-expressions, body posture, and voice tone,[32] which China is using to identify suspicious individuals or emotional states associated with agitation or dissent. This tech is deeply embedded in its public security apparatus, which means that yes, they are using it to identify threats against the state, but with China, that also means using it to surveil their own citizens.[33]

Social media influence operations powered by Behavioral and Cognitive AI is an increasingly used real-world application and now referred to as cognitive warfare. Instead of physical combat, missile targeting or warning systems, or security, cognitive warfare is an evolved type of propaganda defense, aiming to influence, disrupt, and even destabilize societies through manipulation of belief and perception. Behavioral AI is used by tracking, predicting, and even influencing human behavior on digital platforms. It analyzes massive datasets from online activity, things like browsing patterns, sentiment,

---

[30] Hunder, M. and Balmforth, T. (2025). Auterion says it will provide Ukraine with 33,000 AI drone guidance kits. *Reuters* (28 July). https://www.reuters.com/business/aerospace-defense/auterion-says-it-will-provide-ukraine-with-33000-ai-drone-guidance-kits-2025-07-28.

[31] Bergengruen, V. (2023). Ukraine's 'secret weapon' against Russia is a controversial U.S. tech company. *TIME Magazine* (14 November). https://time.com/6334176/ukraine-clearview-ai-russia.

[32] Teh, C. (2021). 'Every smile you fake' — an AI emotion-recognition system can assess how 'happy' China's workers are in the office. *Business Insider* (15 June). https://www.businessinsider.com/ai-emotion-recognition-system-tracks-how-happy-chinas-workers-are-2021-6.

[33] Lanlan H. and Xiaoyi, L. (2021). China leads in emotion recognition tech, reinforces privacy rules to tackle abuse. *Global Times* (3 March). https://www.globaltimes.cn/page/202103/1217212.shtml.

likes, and then creates precise psychological profiles of individuals or groups. The profiles are used to tailor content that elicits specific emotions from the targeted people or groups, be it fear, anger, trust, etc., ultimately pushing people toward certain attitudes and beliefs.

During the 2016 U.S. presidential election, a Russian state-linked organization used this cognitive warfare tactic at the request of the Kremlin to harm then-candidate Hillary Clinton's chances of success.[34] Similar strategies have been used in the ongoing Ukraine–Russia war and Israel–Gaza conflict, both of which have seen AI-driven psychological operations attempting to demoralize populations, discredit leadership, and manipulate international public opinion.

Another emerging application of Behavioral in a similar realm is for detecting deepfakes. In early 2025, *The Wall Street Journal* ran a chilling headline: "The Panicked Voice on the Phone Sounded Like Her Daughter. It Wasn't."[35] A mother received a call from someone who sounded exactly like her child, begging for help, pleading through tears. It was urgent, emotional—and entirely fake. The voice was a deepfake, synthetically generated using AI. The call was part of a scam, but it could have easily triggered much worse. Stories like this aren't science fiction anymore. Audio deepfakes are now good enough to fool not just parents, but voice authentication systems, coworkers, and even military command chains. And with voice models like ElevenLabs, MetaVoice, and OpenVoice openly available, the barrier to entry is alarmingly low. We are officially in the era where hearing is no longer believing.[36]

Most deepfake detection models today are built around the idea of vocal biomarkers. These are subtle physical characteristics that naturally occur when we speak: micro-pauses, pitch variation, throat resonance, the slight jitter in frequency caused by emotion or breathing. Our real voices are full of these artifacts, and the assumption has been that synthetic voices just can't replicate them consistently. The problem is, deepfake generation models are evolving fast. What used to be vocoder artifacts or synthetic "tells" are now being patched.

[34] 116th Congress (2019). Senate Intel Committee releases bipartisan report on Russia's use of social media. Press release (8 October). https://www.intelligence.senate.gov/2019/10/08/press-senate-intel-committee-releases-bipartisan-report-russia-e2-80-99s-use-social-media.

[35] Jargon, J. (2025) The panicked voice on the phone sounded like her daughter. It wasn't. *The Wall Street Journal* (5 April). https://www.wsj.com/tech/personal-tech/the-panicked-voice-on-the-phone-sounded-like-her-daughter-it-wasnt-8d04cbc1?mod=djemTECH.

[36] Gujral, R. (2025). AI vs AI: The arms race for authenticity. LinkedIn. https://www.linkedin.com/pulse/ai-vs-arms-race-authenticity-rana-gujral-ciyzc.

Models like HiFi-GAN, WaveGlow, ParallelWaveGAN, or the newer NaturalSpeech 2 are capable of synthesizing speech with very few detectable artifacts.

This is very similar to the diamond industry, where lab-grown diamonds are chemically identical to those formed naturally, but jewelers have traditionally distinguished them through two features: serial numbers embedded by some mines and natural imperfections called inclusions. Inclusions, or tiny flaws, signal that a diamond is naturally formed; when a diamond appears too flawless, it's often suspected to be synthetic. However, manufacturers can now add artificial inclusions during the lab-growing process, rendering these traditional indicators ineffective. This is a net positive for the world, as the diamond industry as a whole faces ethical concerns related to conflict, labor practices, and environmental impact, and blood diamonds, which are diamonds mined in war zones and sold to finance armed conflicts against legitimate governments, are still a significant issue. Lab-grown diamonds offer an ethical alternative to mined diamonds that are virtually indistinguishable to the naked eye, with only specialized lab tests able to differentiate.

This parallels developments in deepfake voice technology but in a more negative manner, where AI-generated speech, originally detectable due to the absence of natural vocal "imperfections" such as shimmer or tonal variation, can now include synthetic versions of these markers. With convolutional neural networks capable of embedding these features into generated audio, distinguishing real from fake becomes increasingly difficult, raising complex questions about authenticity, perception, and the nature of "realness" in an era of indistinguishable replication.

This is where behavioral mapping comes in—a more nuanced approach that doesn't just analyze vocal signals at the surface level but attempts to model how a specific person speaks over time. Instead of relying solely on static biometrics or visual forensics, these systems analyze subtle patterns in speech, facial micro-expressions, vocal cadence, and interaction styles. Each individual exhibits unique behavioral signatures that are difficult to replicate precisely. By comparing real-time behavioral data with authenticated baseline profiles, Behavioral AI can detect deviations that indicate potential manipulation. This method enables dynamic and context-aware verification and serves as a critical tool in sectors where identity authenticity and information integrity are essential, such as media, finance, and national security. Temporal modeling takes it to the next level, identifying when behaviors occur and how they change over time, as behavioral signals are most reliable across time. Detectors model prosody trajectories, micro expression timing, and co-articulation patterns across sentences. Shifts in these temporal rhythms reveal synthesis even when single frames or single audio windows look perfect.

Behavioral Signals provides technology that aids in deepfake detection and distinguishing between natural and synthetic voice and has the ability to detect voice variances in real-time speaking engagements. We also use Emotion AI and Behavioral AI to assess a candidate or threat before polygraph testing, benchmarking their natural cognitive state against pressure test conversations and polygraph tests to detect inconsistencies or deceptions. Finally, our technology has the capability to use vocal biomarkers to predict human behavior like intent (do harm or no harm), detect stress or duress, detect fraud, detect "control" and trustworthiness, providing real-time feedback to operators, improving their communication skills, empathy, and crisis management capabilities.[37]

As excited as I am about all the ways AI is evolving and expanding the defense sector, I would be remiss if I didn't disclose that this technology can, and is, being used in ways that violate human rights. We have already touched on deepfakes and cognitive warfare and the fact that these are created and used not just by ordinary people but by powerful governments, but several other areas cause alarm for the people who are not in power.

AI systems that use Emotion AI to identify threats or facial recognition software[38] rely on unverified indicators like specific micro-expression patterns and have routinely been criticized for scientific unreliability. These systems are prone to racial and cultural biases and there are broader privacy concerns, such as in the case of the Chinese government, that systems are not just being used to keep citizens safe but instead to target civilians or political enemies specifically.

The Future Attribute Screening Technology (FAST) program, originally titled Project Hostile Intent, was designed to detect malicious intent through physiological and behavioral screening and was planned to be used in airport and border security. However, the system faced significant controversy and criticism and is still an "ongoing research project" rather than being a fully deployed operational

---

[37] Behavioral Signals. Leveraging Emotion AI to detect, track, monitor, and model behaviors. https://behavioralsignals.com/defense-intelligence.

[38] Which does not utilize Emotion, Behavioral, or Cognitive AI.

system. The system measures pulse rate, skin temperature, breathing, facial expression, body movement, pupil dilation, and other physiological behavioral patterns. Researchers like Tom Ormerod of the Investigative Expertise Unit at the UK's Lancaster University flagged the system as capable of major false positives because situations like flying and traveling at an airport, or crossing a border, can be hard and anxiety provoking for even the most basic, harmless people. These situations are naturally stressful on top of the fact that emotion is deeply cultural and personal. Misinterpretation can have serious consequences for individuals and families. Whether this is an issue that governments care about is another debate.

Privacy is a major issue, which we will discuss more in further chapters, but who owns the emotional and behavioral data of the Emotion AI and Behavioral AI systems? What exactly is being recorded, and for how long? This data is deeply personal, often unconscious, and difficult to anonymize. There are major issues around consent and mass surveillance that governments often bypass through laws like the USA PATRIOT Act which rapidly expanded governmental authority to conduct surveillance on domestic and international communications in the wake of 9/11 in 2001. Governments cite a need for security and safety for citizens, but we know that these same laws cause harm to citizens who are caught in the crossfire.

An example is that the Muslim American community faced heightened surveillance and systemic harassment by federal agencies, particularly the FBI, after the passing of the USA PATRIOT Act. Undercover informants were placed in mosques without suspicion of criminal activity, solely to monitor Muslims based on their religion. This led to significant fear, community distrust, and violations of civil liberties. Imam Yassir Fazaga and others brought a lawsuit to the Supreme Court in 2022, but the Court ruled in favor of the FBI.[39]

The best defense is a strong offense, and while anyone who plays or watches sports knows this to be true, unfortunately peacemaking and international relations are vastly considered separate from defense and have different goals. That said, Strategic Headwinds, a Behavioral and Cognitive AI project from the Center for Strategic and International Studies' (CSIS) Future Labs and working with the U.S. National Security Council (NSC), aims to be helpful to Western

[39] Mansoor, S. (2021). Who else is spying on me?' Muslim Americans bring the fight against surveillance to the Supreme Court. *TIME Magazine* (16 September). https://time.com/6097712/muslim-american-surveillance-supreme-court-sept-11.

diplomats in securing a peace deal between Ukraine and Russia. The Ukraine–Russia Peace Agreement Simulator[40] analyzed information like media analyses of issues at stake in the Russia–Ukraine war, answers provided by specialists to a questionnaire about the relative values of potential negotiation trade-offs, and a database of 374 peace agreements and ceasefires.

On a scale of one to ten, the software scores the likelihood that the components of the created peace deal would be satisfactory to not only Russia and Ukraine but the U.S. and Europe, based on four areas: territory and sovereignty, security arrangements, justice and accountability, and economic conditions.[41] This model is not perfect, however, and in tests that Future Labs did on different AI programs aimed at peacekeeping, one model, Llama, opted to use force 45% of the time in the test runs for peace negotiations.

Education and Personalized Learning AI

Meant to detect, assess, and nurture students' emotions, Emotion AI and Behavioral AI is being used in education and personalized learning settings, working to optimize learning styles and help youth. We know that emotions significantly impact things like attention, memory encoding and retrieval, cognitive processing capacity, and motivation to learn, so companies like SensEmo and AutoTutor are using Emotion AI to help students. SensEmo has smartwatch signals that can infer students' emotional states and provide real-time feedback to instructors. In initial trials, it achieved approximately 88.9% accuracy and improved quiz scores by around 40%.[42] AutoTutor is a conversational tutoring agent that integrates emotion recognition alongside cognitive assessment, allowing it to adapt dialogue for instructors based on inferred emotional states in addition to student

[40] Center for Strategic and International Studies (CSIS), Futures Lab, "Strategic Headwinds: Understanding the Forces Shaping Ukraine's Path to Peace," which includes the Ukraine–Russia Peace Agreement Simulator and describes its underlying approach and inputs (including expert elicitation and analysis of negotiation items and tradeoffs). https://www.csis.org/programs/futures-lab/projects/strategic-headwinds-understanding-forces-shaping-ukraines-path-peace.

[41] *The Economist* (2025). AI models could help negotiators secure peace deals. *The Economist* (16 April). https://www.economist.com/science-and-technology/2025/04/16/ai-models-could-help-negotiators-secure-peace-deals.

[42] Choksi, K., Chen, H., Joshi, K., et al. (2024). SensEmo: Enabling affective learning through real-time emotion recognition with smartwatches. In *2024 IEEE 21st International Conference on Mobile Ad-Hoc and Smart Systems (MASS)* (pp. 453–459). IEEE. https://arxiv.org/abs/2407.09911.

understanding.[43] Adaptive learning platforms from companies like Fishtree, Knewton Alta, and Realizeit are powered by Behavioral AI. They continuously assess learner behavior and tailor learning trajectories and lesson plans based on how students interact with the system and their particular learning style. For neurodivergent individuals, instructors are using AI models and training to create empathy-rich educational tools that help students with conditions like autism, ADHD, or dyslexia. For example, AI can detect when a student starts feeling overwhelmed, disengaged, or even confused through voice, facial, and behavioral clues and then prompt the instructor to adjust content delivery, pacing, or switch to a different subject entirely.

Criminal Justice and Law Enforcement

Behavioral AI is actively being explored in criminal justice and law enforcement verticals, particularly in lie detection and interrogation settings.[44] The company Converus created the EyeDetect+, which is currently used by U.S. and international law enforcement agencies for screening officer candidates as well as conducting internal investigations. EyeDetect+ is an eye-tracking software that analyzes eye movements, pupil dilation, and reaction times to infer deceptive behavior and lie detection. Other startups, such as Silent Talker, CyberQ, Arche AI, and Deceptio.AI, are developing real-time deception detection tools for integration into interviews or virtual courtrooms that will analyze facial microexpressions, speech hesitation, and vocal tone to assess credibility. These AI tools are meant to offer automated behavioral insight to supplement human judgment.

Marketing, User Experience, and Automotive Safety

One vertical deploying really fun and creative Emotion, Behavioral, and Cognitive AI is in marketing and user experience. One company leading in this vertical, and who was one of the pioneering companies in Emotion AI, is Affectiva. Their technology was created in the MIT

[43] Graesser, A.C.; VanLehn, K.; Rosé, C.P., et al. (2001). Intelligent tutoring systems with conversational dialogue. *AI Magazine* 22(4). https://doi.org/10.1609/aimag.v22i4.1591.

[44] Its use is highly limited and under scrutiny.

Media Lab when cofounders Dr. Rana el Kaliouby and Dr. Rosalind Picard developed a device to help autistic children better communicate by interpreting human emotions. The company went on to specialize in marketing, with their technology deployed by major brands and media firms to assess audience engagement, emotional valence, and cognitive states during ad testing and content evaluation. In 2021 they were acquired by SmartEye, a Swedish company that had developed Human Insight AI, technology that analyzes, supports, and predicts human behavior in complex environments. For example, their technology was leveraged by agencies like Kantar and MarketCast to measure emotional engagement in video ads and trailers. One study involving a major action movie franchise found that tracking emotional moments like surprise and engagement helped optimize marketing assets, like geography, category, and media length, leading to higher resonance across audiences and improved campaign performance.[45,46]

Affectiva and SmartEye also work in the automobile safety vertical for in-cabin monitoring to assess driver and occupant emotional and cognitive states, including drowsiness, distraction, anger, and focused attention. This technology can be integrated into vehicles for real-time driver-state analysis, triggering adaptive alerts or actions such as visual or auditory warnings, climate adjustments, or transition to autonomous driving if impairment is detected.

While this is not a comprehensive list of applications, this is a thorough overview of how these AI technologies are being deployed across diverse sectors, impacting everything from safety to entertainment to health. Looking ahead, ongoing research into multimodal foundation models, AI alignment, hybrid human–AI cognition, and neurosymbolic AI promise to push AI toward unprecedented levels of reasoning, planning, empathy, and adaptivity. What this looks like across these verticals and others has yet to be seen, but we have a pretty good idea of what that will be like. The next wave is multimodal and long context. Video tokenizers and audio language models will let agents reason over hours of interaction, plans, and evidence. With longer memory and better tools, the question is no longer what the model can say, but what it should do, when to stop, and when to ask for help.

---

[45] McManus, A. (2024) Enhancing consumer insights: Affectiva Media Analytics with Emotion AI and facial coding. *Smarteye* blog (14 June). https://www.smarteye.se/blog/enhancing-consumer-insights-affectiva-media-analytics-with-emotion-ai-and-facial-coding.

[46] McManus, A. (2024). 4 cases of facial coding in film testing: what makes viewers feel? *iMotions A/S* blog (5 August). https://imotions.com/blog/facial-coding-films.

Looking Ahead

As research and applications progress, we continue to explore critical questions around causality, ethics, consciousness, and collaborative cognition. Ultimately, AI's future trajectory includes potential breakthroughs in machine sentience, adaptive learning, autonomous improvement, and the profound philosophical implications accompanying the rise of artificial general intelligence (AGI) and then artificial superintelligence (ASI).

CHAPTER

4

The Emergence of Artificial General Intelligence (AGI)

Even if you didn't know that much about AI before picking up this book, artificial general intelligence (AGI) is likely a term that you have heard somewhere on social media or television. The term was coined by Mark Gubrud in 1997, but it was popularized by Ben Goertzel and Cassio Pennachin[1] when they published their 2007 book by the same title, *Artificial General Intelligence* (Springer),[2,3] which helped shape the modern understanding of the term.

Goertzel believes that AGI means an AI entity capable of understanding and problem-solving with structural cognition that goes beyond pattern recognition. Where OpenAI's definition of AGI leaves us thinking that we may have already achieved AGI, the truth is that many of those tasks can be completed successfully by LLMs because pattern recognition is their core function. Goertzel and Legg distinguished AGI and the term from narrower forms of AI, emphasizing the potential for a truly general, human-level intelligence in machines.

Elon Musk[4] frequently talks about AGI in a way that creates visions in our minds of robots walking around as our personal assistants. But what is AGI exactly and when will we achieve it? The truth is much disagreement exists on both of those counts, not just based on

---

[1] AI entrepreneur and editor on the book.

[2] Goertzel, B. and Pennachin, C. (eds.) (2007). *Artificial General Intelligence.* Springer Berlin, Heidelberg. `https://doi.org/10.1007/978-3-540-68677-4`.

[3] Goertzel credits Shane Legg, ML researcher, for choosing the title for the book.

[4] International businessman and entrepreneur and the wealthiest person in the world as of 2021 up to the time of publication.

varying levels of hope among the scientists working on it but because AGI means different things to different people.

The Many Interpretations of AGI

There are myriad definitions of AGI out there. OpenAI[5] defines it according to business logic: AGI is a technology that can perform most economically valuable tasks better than humans can do. It suits them to use this definition. Demis Hassabis[6] believes AGI to be achieved when AI can not only match human pattern recognition, but has reasoning, creativity, and cognitive adaptability as well.

Jeff Clune[7] defines AGI as an AI system with the capability to autonomously learn and adapt to perform the full range of human cognitive tasks but also has the capability to evolve over time, as do biological organisms. Mark Zuckerberg has said he does not have a one-sentence definition. DeepMind instead proposes "levels of AGI." Their ICML[8] framework grades progress from "Emerging" to "Superhuman," which helps operationalize milestones rather than argue semantics.

Definitions also vary based on which discipline you're coming from. Engineers define AGI as an AI system that can perform any cognitive task that a human can do. Philosophers believe AGI will be achieved when there is an understanding, self-awareness, and even consciousness within AI. If you're a layperson interested in AI, you might think that AGI is achieved when AI can pass a universal Turing Test[9] that shows it is at or above human-level cognition and can reason as well as interact in the real world without markers, or any obvious giveaways that the system is not actually human.

These differing perspectives show not only how varied our definitions of AGI are, but also how they shape expectations about *when* we might see it realized. There is also no definitive timeline here. Hassabis often says five to ten years, Sam Altman says OpenAI already knows

---

[5] American AI company and creator of ChatGPT.

[6] British artificial intelligence researcher and entrepreneur; CEO of DeepMind Technologies.

[7] Canadian computer scientist and researcher; founding member of Uber AI Labs.

[8] Morris, M.R., Sohl-Dickstein, J., Fiedel, N. et al. (2024). Levels of AGI for operationalizing progress on the path to AGI. Proceedings of ICML 2024. `https://doi.org/10.48550/arXiv.2311.02462`.

[9] A test of a machine's ability to exhibit intelligent behavior equivalent to that of a human.

how to build AGI, Musk has floated 2029, and Clune thinks there's roughly a 30% chance that AGI is achieved by 2030.

Yann LeCun[10] believes that AGI is decades away and will not be the result of a specific breakthrough but years of building upon current and future projects. He has also argued that the term itself is misleading, due to all the various definitions, and suggests using a term like "human-level AI" might bring everyone to the same definition. However, he goes further to point out that even human intelligence is highly specialized and that an AI matching it would still be narrow, not truly general.

Defining Human Intelligence

How do we humans define intelligence? The Theory of Multiple Intelligences[11] shows us that there are several distinct types of ways to be intelligent beyond traditional academic measures, including linguistic, logical, mathematical, spatial, musical, bodily-kinesthetic, interpersonal, intrapersonal, naturalistic, and potentially existential intelligence. People possess different strengths and ways of being smart, rather than there existing a single, unified intelligence.

Society often refers to people as being street-smart versus book smart. The insinuation is that people can be considered to be smart in different ways. If someone can speak seven languages but is not very strong in mathematics, are they not still considered intelligent? A PhD candidate in cell biology has different intellectual strengths and understandings than an Army general, but each is intelligent in their own way. Levels of intelligence differ when you're speaking about a four-year-old versus a 24-year-old, but that doesn't mean that one is less intelligent than the other. We consider their abilities for understanding the world.

Additionally, Intelligence Quotient (IQ) measures cognitive abilities like logic and reasoning, whereas Emotional Quotient (EQ) gauges a person's ability to understand and manage emotions. Managing emotions is considered to be extremely important in present-day society, and although EQ doesn't have the word "intelligence" in it, most people consider someone who has a high EQ to be intelligent.

[10] French-American computer scientist working primarily in the fields of machine learning, computer vision, mobile robotics, and computational neuroscience; Chief AI Scientist of Meta.

[11] Proposed in 1983 by Howard Gardner, an American developmental psychologist.

Thus, the spectrum for intelligence is vast, and its definition remains conditional, or fluid, based on what or whom you're discussing. This fluidity persists because people and companies often benefit from definitions of intelligence and AGI, and timelines for achievement, that can be tailored to their needs. Karen Hao[12] writes about this in *Empire of AI* (Penguin Press), which speaks about how all the recent progress made in AI, which is vastly reshaping earth, climate, and humanity as we know it, is based entirely on economic value for a select few.

Ultimately, intelligence when discussing AI is often left undefined outside of the specific companies and labs that are working on their models. "Intelligence" is used interchangeably with "cognition," "speed," or "optimization."

With definitions drifting between capability, economic impact, and social transformation, treating AGI as a single finish line creates confusion. Since we cannot even define AGI properly or agree on a definition, how do we truly know what it is? How do we achieve the goal if we cannot even write the goal down? Timelines are opinions until they are tied to auditable benchmarks. Claims range from "a few years" to "decades," and mostly reflect differing priors about whether scaling will plateau or integrate with new ingredients like causality, embodiment, and memory.

With all the fantastic minds working on this, I believe we can do better. I also feel that the current debate is, at best, immature, and that we are asking the wrong questions. I suggest we move the conversation from "What is AGI?" to "Which AGI, and measured how?" Currently, general agreement exists that there are three areas or definitions related to AGI:

1. Capabilities: reasoning, planning, causal modeling, memory persistence, tool use, interpretability, etc.
2. Embodiment and Grounding: sensorimotor competence in either rich simulation or the physical world
3. Reliability and Economics: contribution to economically valuable work

The fourth area is Artificial General Experience (AGE). This area is less known as I recently coined the term, but I believe AGE is vital to this conversation. AGE is important because intelligence without

[12] American journalist, NYT best-selling author, and TIME100 honoree, specializing in the intersection of AI and society.

experience is hollow. Until we measure the path to intelligence and the path to experience side by side, "AGI when" is the wrong question.

The Love of Wisdom

Goertzel, and others like me and Bach, believe that AGI development is both scientific and philosophical. The world wants to create a machine that meets or surpasses the intelligence of humans, but as I discussed earlier in this chapter, the key to human intelligence is emotion. Our first stabs at machines that learn and replicate human emotion are Emotion AI and Behavioral AI, but those are simply steppingstones on the road to AGI.

Currently, the AIs that are being built, for purposes in marketing, defense, and healthcare (see Chapter 3), are powerful but unaware. AI will continue to grow in power and "intelligence" as companies and scientists race to the finish line of achieving AGI and, potentially, artificial super intelligence (ASI).[13] And, unless we do it correctly, these AIs will become extremely powerful systems that will be participating in the world without true emotion and understanding of their relationship to it. Without understanding, there can be no common ground between AI and humans, or even AI and the planet.

This is the mission of Bach's latest venture, The California Institute for Machine Consciousness (CIMC),[14,15] established on the belief that the key to a flourishing future lies in developing AI agents that understand human subjective experience, coexist with other agents, and navigate ethical agency within dynamic, multi-agent systems. Bach also believes that AGI and machine consciousness will be here within the next 3–5 years.

He states that in order to develop safe and ethical agents, we must better understand consciousness, which is central to understanding the human mind yet remains largely neglected by current scientific fields. Psychology, neuroscience, and AI mostly sidestep the issue, focusing on outputs rather than theories of consciousness itself. Without comprehensive models or metaphysics to address it, a divide persists between scientific approaches and lived human experience, limiting progress in both AI and human development.

[13] We will discuss this more in Chapter 7, "The Road Ahead: What Lies Beyond AGI."

[14] I serve as a strategic advisor with the CIMC.

[15] See `https://cimc.ai/#`.

The CIMC proposes that consciousness is a self-organizing algorithm implemented on biological brains, one that can be understood and replicated in nonbiological systems. It aims to create a synthetic architecture for minds based on modular cognition, coherence, and internal simulation. Human consciousness, Bach suggests, may ultimately be understood as software patterned computation running on a particular substrate. If so, then achieving consciousness in machines would be architectural, not speculative. Or rather, the challenge is one of engineering design rather than philosophical possibility.

Without this understanding, we run the risk of creating AGI that cannot empathize or sympathize with humans, which makes for a dangerous possibility.

Human Experience Defines Human Intelligence

Bach's theory aligns with my own that yes, we are working to achieve artificial general intelligence, but what we're truly working to replicate is the human experience through artificial general experience—AGE. At its root, I believe that intelligence is the ability to adapt meaningfully to the world: to interpret, to prioritize, to decide with relevance. And all these functions emerge from experience. Experience is not a byproduct of intelligence. It is its foundation. AGE introduces a new lens for thinking about the future of AI. It is not a separate step from AGI, but rather a more precise way of understanding what we are already pursuing.[16]

AGE centers on this replication of human experience, encompassing memory, emotion, perception, language, narrative, self-awareness, embodiment, and the internal coherence that gives rise to subjective life. AGI focuses on intelligence and problem-solving, but intelligence without experience is hollow. AGE shifts attention to the core: it is not just about solving problems. It is about generating presence, awareness, feeling, and above all, intuition.

The foundation of AGE draws on the idea that experience does not require a biological substrate. It may arise from the right kind of computational organization. Embodiment and context are vital but not necessarily exclusive to organic systems. Shared perception environments between humans and AI offer early signs of symbiotic experience. In some views, human minds navigate a computational

[16] Gujral, R. (2025). The AI instinct: Beyond intelligence toward Artificial General Experience (AGE). LinkedIn. https://www.linkedin.com/pulse/ai-instinct-beyond-intelligence-toward-artificial-general-rana-gujral-utcmc.

universe by interpreting patterns, compressing reality into coherent narrative.

Models of Consciousness

Stephen Wolfram[17] introduces the idea of the *ruliad*, the entangled limit of all possible computations.[18] Human observers, embedded in this vast computational structure, sample specific pockets of reducibility to generate the world they perceive. In this light, replicating human experience is a matter of recreating the sampling and interpretive structures of the mind itself.

Bach offers a model of consciousness as a coherence-generating operator. Consciousness, in his view, is not a fixed object but a dynamic process that aligns perception, memory, attention, and emotion across time. He describes this flow as a bubble of nowness, a continuous sense of self constructed by second-order awareness. This view is embodied in his work on the CIMC.

Sir Roger Penrose[19] and Stuart Hameroff[20] think that consciousness arises from quantum processes in microtubules inside neurons, setting them apart from most neuroscientists who think consciousness arises from networks of neurons firing in the brain using ordinary biochemistry.[21]

The Penrose–Hameroff microtubule hypothesis, also known as the *Orchestrated Objective Reduction (Orch-Or)*, states that consciousness depends on quantum effects inside microtubules, which are tiny structural components within neurons.[22]

Their theory suggests (i) microtubules can support delicate quantum states, (ii) these quantum processes collapse in a way that is tied to fundamental physics, not just biology, (iii) when orchestrated across

[17] British-American computer scientist, physicist, and businessman.

[18] Wolfram, S. (2021). The concept of the Ruliad. *Stephen Wolfram: Writings* (blog). `https://writings.stephenwolfram.com/2021/11/the-concept-of-the-ruliad.`

[19] English mathematician, mathematical physicist, philosopher of science, and Nobel Laureate in Physics.

[20] American anesthesiologist and professor at the University of Arizona known for his studies of consciousness.

[21] Hameroff, S. and Penrose, R. (2014). Consciousness in the universe: A review of the 'Orch OR' theory. *Physics of Life Reviews*, 11(1): pp. 39–78. `https://doi.org/10.1016/j.plrev.2013.08.002.`

[22] Hameroff, S. and Penrose, R. (1996). Orchestrated objective reduction of quantum coherence in brain microtubules: The 'Orch OR' model for consciousness. *Mathematics and Computers in Simulation*, 40(3–4): pp. 453–480. `https://doi.org/10.1016/0378-4754(96)80476-9.`

many neurons, these collapses create moments of conscious awareness. They argue that classical physics can't fully explain consciousness for a few reasons, including that neural firing alone doesn't capture subjective experience (qualia). Quantum mechanics allows for non-local connections and probabilistic outcomes that resemble the fluidity of thought.

Critics of this hypothesis argue that the brain is warm and wet, which classically destroys fragile quantum states almost instantly, and little experimental evidence exists that microtubules can maintain quantum coherence long enough to matter for that. However, even if Orch-OR is wrong, it pushes science to ask whether quantum physics and consciousness may be linked. It keeps alive the idea that human awareness may not be fully explainable with classical computation alone and that new physics could be involved.

Suppose consciousness does touch the quantum substrate. Suppose intention and attention can bias which possibilities become actual. Then the mind is not only modeling the world. It is also a small part of how the world comes to be. That claim has enormous reach. It touches ethics, therapy, creativity, and the way we build human AI systems. The right response is not to dismiss it. The right response is to set a higher bar for proof while also asking what we would change in practice if it were true.

Nature *does* use quantum effects in some settings. Energy moves through photosynthetic complexes with signatures that look quantum. Migratory birds likely sense magnetic fields through a quantum radical-pair mechanism in proteins. These cases show that biology can exploit delicate physics under the right constraints. They do not show that human brains compute with qubits, but I believe this is an area to watch.

I find the Penrose–Hameroff microtubule hypothesis fascinating. It is undoubtedly an imaginative proposal. It has a few intriguing hints and many open problems. That said, evidence that links microtubule dynamics to conscious experience in humans is not yet persuasive. If future data proves a decisive role, I'd be the first to celebrate it.

In *The Fabric of Reality* (Viking Adult), David Deutsch[23] introduces the many-worlds interpretation of quantum mechanics, the idea that whenever a quantum event[24] could have multiple outcomes, all of them actually occur in separate, parallel universes. Instead of one

[23] David Elieser Deutsch is a British physicist at the University of Oxford, often described as the "father of quantum computing."

[24] An occurrence that happens at the subatomic level and cannot be explained by the rules of classical physics.

reality, a vast multiverse exists where every possibility plays out. Traditional quantum systems are described by a wavefunction that evolves smoothly according to the Schrödinger equation. Rather than "collapsing" to a single outcome when measured, Deutsch says that the full quantum state continues to evolve, and all outcomes compatible with the wavefunction occur in parallel branches of reality.

He does go on to say that "many-worlds" does not imply one can hop between branches or manifest outcomes by will. Branches do not communicate once decoherence separates them. This is personally disappointing, as I would like to believe that we can jump.

Sam Harris[25] has argued that free will is an illusion, and that our decisions are not consciously authored.[26] New findings in neuroscience and cognition suggest that he may be on to something and that we may experience what the brain generates, not what we choose, and that we may not consciously generate our thoughts but instead receive them. Our decisions arise in the brain before we are aware of them. In experiments, brain activity indicating a decision appears milliseconds before we report making that decision.

Benjamin Libet[27] conducted an experiment[28] where he measured the brain's "readiness potential" (a slow buildup of neural activity) before a voluntary finger flex and asked participants to report the moment they first felt the urge to move. He found that the readiness potential began about ~550 milliseconds before movement, and he reported urge appeared about ~200 milliseconds before movement. So neural activity preceded the conscious intention by ~300–350 milliseconds. This paper[29] is the classic citation for "the brain decides before you do."

While many believe this experiment to suggest that our sense of conscious choice may be more of an illusion than a genuine act of will, Libet himself did not interpret it as evidence that our decisions are predetermined. He believed that the experiment shows that the "tendency" to press a button may be building up for 500 milliseconds, but the conscious mind retains the right to veto any action at the last moment.

---

[25] American philosopher, neuroscientist, author, and podcast host.

[26] Harris, Sam. (2012). *Free Will*. Free Press.

[27] American neuroscientist who pioneered the field of human consciousness.

[28] Libet, B. (1999). Do we have free will? *Journal of Consciousness Studies* 6(8–9): pp. 47–57. `https://spot.colorado.edu/~tooley/Benjamin%20Libet.pdf`.

[29] Libet, B., Gleason, C.A., Wright, E.W. et al. (1983). Time of conscious intention to act in relation to onset of cerebral activity (readiness-potential): The unconscious initiation of a freely voluntary act. *Brain* 106(3): pp. 623–642.

A similar experiment[30] by John-Dylan Haynes[31] and his team showed that patterns in the prefrontal and parietal cortex could predict a left versus right button press up to ~7–10 seconds before people reported deciding. Later intracranial recordings found single-neuron signatures building toward a self-initiated act hundreds of microseconds to seconds before awareness.

Critics have contended that these patterns could simply be an indication of premeditation or a thought process in the brain: we think about a decision, choose a decision unconsciously, and then make the decision consciously. Aaron Schurger,[32] Jacobo Sitt, and Stanislas Dehaene[33] posit that the "readiness potential" can be explained by stochastic accumulation toward a threshold, which softens the claim that unconscious brain activity determines the decision.

Overall, the data here supports a modest claim. Much of the work for simple actions happens before we know it, and awareness lags the underlying processes. That fits both a standard computational view and Sam Harris' filtered consciousness lens. However, it does not force the conclusion that thoughts are received from outside the system. In the end this remains unsettled. But I do want to explore this area further, because of the relevance to the overall conversation regarding AGI and eventually ASI.

From Human Consciousness to Artificial Intelligence

Building on the original idea covered in Chapter 2, human behaviors might be less about free, autonomous thought and more about evolutionary programming. If so, everything we do is simply a script of biology and environment playing out. We feel like we're in control but may not be, and even spiritual practices like meditation don't give us control; they let us observe the lack of it.

Extending this perspective, the evolution of intelligence, including the emergence of AI, seems less like a choice and more like an

[30] Soon, C.S., Brass, M., Heinze, H.J., and Haynes, J.D. (2008). Unconscious determinants of free decisions in the human brain. *Nature Neuroscience* 11(5): pp. 543–545. https://doi.org/10.1038/nn.2112.

[31] British-German brain researcher.

[32] American cognitive neuroscientist.

[33] Schurger, A., Sitt, J.D., and Dehaene, S. (2012). An accumulator model for spontaneous neural activity prior to self-initiated movement. *Proceedings of the National Academy of Sciences of the United States of America* 109(42): pp. E2904–E2913. https://doi.org/10.1073/pnas.1210467109.

inevitability. Perhaps humanity is on a path toward becoming or creating a new species, not because we decide to, but because it is written into the trajectory of evolution itself. The lines we draw between "human," "good," and "bad" are social constructs, temporary and shifting, while ultimately everything simply is. From this perspective, our minds, our actions, and even the future of intelligence may be less about will and more about destiny.

Tying back to Harris's theory, if the self is a post-hoc construct, and consciousness is a product of unconscious processing, then replicating experience means reconstructing those underlying generative mechanisms. Consciousness becomes less about agency and more about effect.

Elan Barenholtz[34] provides a complementary insight into the nature of cognition. He suggests[35] that human language, like LLMs, is autoregressive in nature, meaning that language is not inherently tied to external reality but functions as a self-contained predictive process. Critics often reduce LLMs to predicting the next word, but this is precisely how human brains handle language, through pattern recognition and prediction rather than abstract concept-mapping. In this sense, each person's brain operates like a personal-scale LLM. Barenholtz builds his case from a combination of empirical insights and philosophical reflection.

His central premise is that language models like GPT have shown us something surprising about language. These models achieve impressive linguistic fluency using only next-token prediction. They are trained entirely on text. They do not perceive or interact with the physical world. Yet they generate coherent language that rivals or exceeds human ability. The startling implication is that language may never have required reference to the world in the first place.

If autoregression is enough, then perhaps language is fundamentally autoregressive in humans too. This is not a metaphor, Barenholtz posits it is a literal description of the underlying mechanism. What seems like intelligence in LLMs, he suggests, is not mimicry but convergence. LLMs are not faking what humans do. They are doing what humans do, because what we do is not what we thought.[36]

---

[34] American professor and researcher specializing in using behavioral and embedded computational approaches to study the brain and behavior.

[35] Barenholtz, E. (2025). Language without meaning: How LLMs exposed our biggest illusion. *Theories of Everything with Curt Jaimungal*. `https://www.youtube.com/watch?v=A36OumnSrWY`.

[36] Gujral R. (2025). LLMs are doing what you do: The hidden architecture of thought. LinkedIn. `https://www.linkedin.com/pulse/llms-doing-what-you-do-hidden-architecture-thought-rana-gujral-2fx0c`.

The idea that words refer to things in the world has long been foundational in both linguistics and philosophy. But Barenholtz argues that this is a mischaracterization. In both humans and machines, the word *red* does not carry perceptual content. It simply exists in relation to other tokens like *color, blood,* and *apple.* It functions entirely within the topology of language. Meaning, in this view, emerges not from pointing outward but from internal relational structure.

Language is thus not anchored in perception. It is not about describing sensory inputs or external facts. It is about navigating a high-dimensional space of token relations. This is why a model with no access to visual or tactile experience can still write convincingly about sunsets, heartbreak, or the taste of coffee. These ideas exist as trajectories in language space, not as perceptual referents.

Bach's work complements this theory as he speaks of the possibility of building machine consciousness through computational models, showing that consciousness and language may emerge from the same underlying principles of patterned computation. Thus, LLMs are not fundamentally different from human cognition; they may even reveal how language and consciousness naturally arise in biological systems. I believe that Barenholtz distinguishes the symbolic, language driven, model building system from perceptual cognition. He affirms that we do have qualia,[37] sensory experiences, and physical grounding. But those processes are distinct. They belong to a different subsystem. Language interacts with perception but is not derived from it. The two systems are integrated but autonomous.[38]

The Nature of Reality

I myself have raised philosophical questions about the constructed nature of self and reality, echoing our deeper analysis of consciousness and how both humans and AI interpret information.[39] I speak about this in a TEDx, sharing how our brain creates a reality for us to experience by taking in everything through our senses and editing it for us. We do not see what actually exists, rather what our brain

[37] The lived feel of experience itself; what it's actually like to taste something or see a color or feel a gut-level emotion.

[38] Jaimungal, C. (2025). The theory that shatters language itself. *Curt Jaimungal* Substack (10 June). https://curtjaimungal.substack.com/p/the-theory-that-shatters-language.

[39] Gujral, R. How our mind shapes reality. TEDx Talks (YouTube). https://www.youtube.com/watch?v=gcV1um7kOMw&t.

tells us exists. The world as we know it is an elaborate illusion created by our brain.

The Constructed World

When we think about consciousness, we can frame it in terms of being "inside the box" versus "outside the box" with the box representing the limits of our perception and cognition. We can only experience reality from within. If consciousness is purely metaphysical, then we'll never be able to step outside the box to see the rules that govern it, deeming it inaccessible to science.

If consciousness is computational, as Bach and others propose, then even from within the box we can uncover some of its rules through math, pattern recognition, and scientific modeling.

There might also be ways to briefly glimpse beyond the box through meditation or psychedelics, which can feel like portals that allow us to see dimensions of consciousness that are normally hidden. This ties into evolution, I see consciousness not as something mystical but as a natural phenomenon that emerged from adaptive processes and with the right computational tools could be understood or replicated. I covered this lightly in Chapter 2 when I discussed how psychedelics impact our reality.

Altered States and the Quiet Mind

In May 2025, I flew to the Amazon in Brazil for my first ayahuasca ceremony. I went to study my own mind. Interestingly, my ceremony was less about visions and more about subtraction. To prepare for the ceremony, I had to eliminate alcohol, meat, caffeine, dairy, and other things. Ayahuasca contains a monoamine oxidase inhibitor (MAOI), and this diet helps prevent dangerous interactions with certain substances, particularly those containing tyramine.

The first thing I noticed during and after the ceremony was that the inner commentator went quiet. The grip of me loosened. Neuroscience suggests this state comes from a temporary softening of the default mode network (DMN) and a relaxation of top-down predictions, which lets raw perception come forward. In that quiet, I could see work and life with less narrative distortion.

DMN is a set of midline hubs, especially posterior cingulate and medial prefrontal cortex, that support self-talk, autobiographical memory, and mind wandering. Under classic psychedelics like psilocybin, LSD, and DMT, activity and connectivity within these hubs drop, and the tight coupling between them loosens. At the same time, communication between many networks becomes more globally connected. This pattern tracks the reports of a quieter inner narrator and a lighter sense of self. My experience helped me see why minds need a tunable self model, a control for how much prior belief to apply, and attention that can widen to explore and narrow to execute. It also showed how memory can reconsolidate with new values when noise drops.

Lessons for Cognitive Design

Another profound shift has been in my ability to observe, both myself and others, with a clarity I didn't have before. This isn't just about what happened during the ceremony; weeks later, the change remains. It feels as though new neural pathways were wired, embedding lasting capabilities. I now notice patterns in my own behavior and in others' actions that previously escaped me. This makes me wonder: if ayahuasca can enhance observation in this way, could similar processes allow us to access entirely new kinds of data or perception? Perhaps, over time, such abilities could bring us closer to understanding and even replicating the full scope of cognition and consciousness.

The mind is a prediction machine that balances top-down expectations with bottom-up signals. Psychedelics seem to relax the weight of high-level expectations, which lets raw data speak more loudly. That is why the inner narrator quiets and why perception can feel clearer. Meditation and psychedelics can both quiet the DMN, but the experiences are different. Meditation trains gradual control over attention and narrative through practice. Psychedelics induce a temporary state shift that shows what that quiet can feel like.

The design lesson for AI is similar. These are the same dials we should build into cognitive architectures. Minds need a self model, but they also need a safe way to relax. AI models need a dial for how much prior belief to impose and a way to listen harder to the data when stuck. That is how we move from narrow problem-solving toward systems with presence, perspective, and better judgment. Build systems that can quiet narrative, listen harder, and then speak again with intent.

Another layer to this is the question of what it would mean for AI if consciousness were filtered, not generated? Could an AI ever unfilter its own perceptions, or is it fundamentally locked in whatever architecture we give it? If the brain filters reality for evolutionary utility, AI might need its own "filters" not just to simulate humans but to survive in human spaces. Perhaps the real challenge in AI is not teaching it to think but teaching it what not to think.[40]

[40] We will explore this more in further chapters.

Simulation and Substrate Independence

Another idea that assumes *substrate independence*, the idea that consciousness doesn't depend on neurons specifically but on the patterns of computation, is that of Nick Bostrom's[41] Simulation Argument.[42] Bostrom offers not a claim, but a trilemma, using math and probability by comparing how many humans could exist in "real" history versus how many could exist inside computer simulations. He shares three propositions: (i) almost all civilizations at our level go extinct before reaching posthuman capability, (ii) almost no posthuman civilizations run significant numbers of ancestor simulations, (iii) almost all observers like us live in simulations without knowing it. He argues that at least one of these is true and the probability move is conditional. If future civilizations can and do run vast numbers of ancestor simulations, simulated observers would vastly outnumber biological originals, so by indifference we should think we are probably simulated.

Substrate independence is an assumption that Bostrom discusses to make his argument coherent, but it is not proven. He is primarily concerned with metaphysics and probability, so his thinking is less about how the mind works and more about whether our universe can, or might be, an engineered simulation.

Engineering Experience

On this assumption, simulated beings could be conscious, which aligns with my claim that experience can be engineered if we build the right computational organization.

Together all these perspectives reveal experience as something deeply computational. Not mystical, but modular. Not reducible to one mechanism, but arising from the convergence of many. The essential components of experience such as attention, memory, emotion, and identity can be understood not just as phenomena to admire, but as blueprints to replicate.

In my view, experience is not magic. It is a stack of interacting processes that we can name, measure, and build. Attention selects what to process now. Memory keeps both episodes and facts and updates them when new evidence arrives. Emotion assigns value to events and guides action. Identity ties moments together so there is a felt

[41] Swedish philosopher.

[42] Bostrom, Nick. (2003). Are you living in a computer simulation? *Philosophical Quarterly* 53(211): pp. 243–255. `https://simulation-argument.com/simulation.`

sense of continuity. These parts do not live in isolation. They talk to each other through learned signals and shared control loops. That is why experience looks mysterious from the outside and modular from the inside. This is an engineering frame, not a final theory of mind.

For years I leaned toward hard determinism. The brain felt like a storyteller that arrived late to its own actions. Lately my view has shifted. The stage is still there, the sets are still props, but attention and intention seem to move real pieces of the play. I now hold two ideas at once. Experience is constructed, and within that construction, we can shape outcomes in reliable ways. Classic experiments show neural activity that predicts simple choices a moment before people report deciding, which weakens the claim that conscious intention is always first. A newer reading is that background neural activity accumulates toward a threshold; when it crosses, the action and the feeling of deciding arise together. Conscious will is not a single magic spark. It is a control system layered on top of deeper processes.

As I shared in my TEDx, there is a version of "you create your reality" that does not require mind over matter. Brains are prediction machines. They constantly guess what is out there, and then update those guesses with incoming signals. Priors and attention shape what we perceive, how tightly we grip a story, and which options we even notice. Beliefs and context also change bodies. Placebo and nocebo effects alter pain, hormones, and performance. Coherence between belief, attention, and action often changes outcomes.

Quantum ideas are sometimes invoked to explain perception. The safe takeaway is modest. At microscopic scales, systems exist in superposed possibilities that become effectively classical through interaction with their environment, a process called decoherence. Whether observation by a conscious mind plays a special role is an open philosophical question, not a settled scientific fact. For this book, "manifestation" is better read as cognitive and behavioral coherence rather than a claim about collapsing wavefunctions.[43]

Understanding how perception, attention, memory, and value interact gives us a blueprint for engineering experience. If we want AGI to become AGE, we will need systems that can tune priors, quiet narrative when stuck, reconsolidate memory with updated values, and act with calibrated intention. That is how we move from raw problem solving toward presence and judgment.

Further, AGE is not opposed to AGI. It is a natural extension of it. AGI gives us tools that compute. AGE asks what it would take to

---

[43] Interpretations of quantum mechanics are not empirical facts.

build something that perceives, reflects, and feels. An AGE system might include modules for generating internal narrative and memory. It would need mechanisms for attention and coherence, to provide temporal alignment. Emotional structures would link events to value. A self-model would integrate these elements into continuity. Each part would reflect a thread of the human experience,[44] not just in behavior but in internal life. This is the bridge from AGI to AGE. Intelligence gives the raw ability to solve problems. The experience stack gives relevance, perspective, and continuity.

Multiple Pathways to AGI

Just as a variety of definitions and timelines have been put forth for AGI, multiple approaches are being worked on to reach it. AGI is the new frontier, the Wild West. While LLMs have captured headlines for their capabilities, they represent just one pathway among many being actively pursued. Labs, universities, and private companies across the world are advancing competing and complementary approaches, the diversity reflecting the uncertainty and excitement surrounding this work: no one knows which path will succeed, although many will die on a hill that their way is the correct way or that certain pathways are hurting the overall goal, sending the sector into another AI winter.

A rigorous pathway to AGI begins with accepting that minds are physical processes, not mystical ones, and therefore, in principle, reproducible. This is the core of the Church–Turing–Deutsch principle: any physically possible process, including cognition, can be simulated by a universal computing substrate. That doesn't imply AGI will be easy or inevitable, only that it is not barred by the laws of physics. But replication is not enough. As Deutsch argues, scale and pattern recognition alone cannot yield general intelligence; true AGI must generate good explanations, create hard-to-vary models that work across domains, and improve those explanations and models through criticism and self-correction. In other words, AGI emerges not from brute computation, but from creating and refining universal explanations, just as humans do.

I dissect the pathways being worked on and include arguments for and against each in order to provide a well-rounded view of the state of our progress.

[44] We will discuss this further in future chapters.

Large Language Models (LLMs)

This is most likely the most well-known pathway, as there are a variety of debates ongoing online and at conferences about this pathway. Overall, it bets on making models like GPT larger, more data-rich, and fine-tuned, believing that AGI will emerge with scale. Proponents believe that complex reasoning, creativity, and planning can be approximated by statistical learning from massive, structured, electronic collections of written texts. It is their belief that scaling plus multimodality and tool use are already yielding broad generalization and creation models that plan at high levels, call tools, and quickly transfer across domains. The idea is that language corpora encode a rich prior[45] over the world, so ever-larger models stitched into agents will get us most of the way to AGI.

Altman and OpenAI have been vocal about LLMs as a foundation for AGI and their GPT series embodied this approach. Dario Amodei[46] of Anthropic, creator of the Claude models, emphasizes that LLMs with strong safety and alignment layers are a credible path to AGI.

Overall, LLMs deliver real, versatile capability and have massively accelerated progress, but repeated evidence shows limits in grounding, causality, and reliable step-by-step reasoning. This is a main argument from many in the space who oppose LLMs as the correct pathway to AGI. Gary Marcus[47] is the loudest, most consistent voice saying LLMs are impressive, but the field is betting too much on them. He believes LLMs memorize patterns rather than build causal models of the world, so reliability, grounding, and step-by-step reasoning remain weak. He has repeated this in Senate testimony and interviews that say the current LLM path shows diminishing returns and is not the best route to robust intelligence.

LeCun is another prominent critic, saying that LLMs lack grounded understanding and causal prediction. He believes that without self-supervised predictive models of the physical and social world, you get brittle talkers. According to him, achieving systems with these capabilities will likely take decades, not years. He is highly critical of the notion that merely scaling up large language models will lead to AGI. He believes LLMs demonstrate impressive pattern matching but lack foundational cognitive capabilities necessary for genuine

[45] A foundational belief or understanding.

[46] American artificial intelligence researcher and entrepreneur; co-founder and CEO of Anthropic, formerly of OpenAI.

[47] American psychologist, cognitive scientist, and author, known for his research on the intersection of cognitive psychology, neuroscience, and artificial intelligence.

intelligence. He has also expressed concern that continued hype of LLMs without meaningful progress could trigger another AI winter.

Emily Bender's[48] group warns that LLMs are "stochastic parrots" and that hype hides limits in understanding and truthfulness. This concern does have merit as hallucinations, weak causal reasoning, and brittleness are real issues that LLMs face.

I mentioned Barenholtz's work previously.[49] Barenholtz says LLMs are closer to how humans produce language than critics allow. His work strengthens the case that LLMs capture something real about human linguistic cognition and his points that coherence, syntax, and even inference can emerge from next-token prediction shows LLMs are not "just parrots." However, Barenholtz suggests that LLMs can be autonomous linguistic engines but are insufficient for grounded intelligence.

LLMs are legitimate models of our linguistic engine but they are not the whole mind. Arguments against LLM as the key pathway to AGI are legitimate, but "LLMs are a dead end" is an overstated and incorrect thought. LLMs supply narrative and coherence, while grounding, value, and presence come from perception, memory, and causal loops. Scaling has clearly delivered broad capability jumps and predictable gains under scaling laws, with Chinchilla-style data scaling shaping today's best models. We should treat them as the narrative cortex inside an agent that also learns a world model and acts.[50]

Causal Models

Most machine learning involves correlation, but causal models actually represent cause-and-effect relationships. Instead of predicting outcomes from patterns, causal models try to answer the questions, "What if I intervene?" or "What if things were different?," ultimately teaching AI about why things happen. This model builds on causal graphs,[51] a visual representation using nodes (variables) and directed arrows (edges) to show direct causal relationships between variables, with arrows pointing from cause to effect.

Unlike standard neural nets, these models allow for counterfactual reasoning, intervention simulation, and generalization. In AGI, the hope is that causal reasoning would allow machines to move from

[48] American linguist and professor specializing in computational linguistics and natural language processing.

[49] See the section, "From Human Consciousness to Artificial Intelligence."

[50] More on this in the "Hybrid Systems / World Models" section.

[51] Pioneered by Judea Pearl, Israeli-American computer scientist and philosopher.

statistical pattern recognition to genuine understanding of the world. Proponents believe this model supports robust reasoning across different environments, enables planning and decision-making closer to how humans think, and prevents brittleness in AI systems that reply only on correlation.

causaLens is one company that has already commercialized causal AI, and their Causal Reasoning Lab[52] delivers tools such as causal effect estimation, structural causal modeling, root cause analysis, and transparent decision-making tools. These systems help organizations model and act upon cause-and-effect relationships, improving trust and business outcomes. DeepMind has actively advanced the field as well, treating causal understanding not as a theoretical idea but as a foundational component of robust, general intelligence. In "Robust Agents Learn Causal World Models,[53]" researchers Jonathan Richens[54] and Tom Everitt[55] demonstrate that any AI agent capable of generalizing across different situations inevitably learns an approximate causal model of how its environment works. Agents aren't explicitly programmed with causality, rather, causal representations emerge as necessary for meeting performance guarantees across distributional shifts. This establishes a formal, mathematical link between causal reasoning and generalization in AI.

Critics have their say, arguing that there are issues with scalability, data limitations, and integration challenges. Ultimately, critics say that this model, while powerful, is restricted to mostly academic settings and is not evolved enough to be a true pathway to AGI.

Cognitive Architectures

This pathway includes a framework that attempts to model the human mind by recreating its key structures and processes using software. While LLMs run on vast data, this approach creates individual modules that simulate mental functions: working memory, long-term memory, planning, reasoning, etc. A control loop manages attention and goals,

[52] causaLens. Causal Reasoning Lab. `https://causalens.com/causal-reasoning-lab`.

[53] Richens, J. and Everitt, T. (2024). Robust agents learn causal world models. Published as a conference paper at ICLR 2024. `https://proceedings.iclr.cc/paper_files/paper/2024/file/44a2b9f7bf9aec3f1fa333ad875b0ee0-Paper-Conference.pdf`.

[54] American research scientist in AI safety.

[55] American research scientist.

similar to how humans focus on relevant tasks while ignoring distractions, and learning occurs through interaction.

It aims to mirror the architecture of the brain, not just its outputs, so that AI can achieve general, flexible intelligence similar to that of humans, involving thinking, remembering, reflecting, and explaining.

While LLMs act like a powerful intuition engine, cognitive architectures provide the slots that intuition plus into. While some believe that these systems are better than LLMs, they are better at specific things that LLMs struggle with, giving you a specific mind-shaped layout with working memory, long-term memory, attention, a rule system, and a decision loop. This layout makes persistence, multi-step control, and explanation easier to both design and measure.

ACT-R (Adaptive Control of Thought – Rational)[56] is a cognitive architecture developed at Carnegie Mellon University, which models cognition as interactions among modules for perception, memory, and decision-making. SOAR,[57] built at the University of Michigan, focuses on integrating problem-solving, learning, and planning into a unified framework. After decades of research into existing architectures like ACT-R, SOAR, and Sigma, John Laird,[58] Christian Lebiere,[59] and Paul Rosenbloom[60] created the Common Model of Cognition (CMC)[61] with aims to capture the functional organization of a human mind and they believe it serves as a standard blueprint for building AGI systems that replicate human-like flexible cognition.

Critics of this pathway argue that while it is a powerful abstraction, it oversimplifies the brain's true complexity. LeCun has brought up issues with its scalability, saying that they work well in lab tests but don't show the kind of open-ended learning LLMs and reinforcement systems have. Others say that this architecture is too hand designed, baking in too many human assumptions instead of letting intelligence emerge naturally from data.

---

[56] Anderson, J.R., Bothell, D., Byrne, M.D. et al. (2004). An integrated theory of the mind. American Psychological Association. `https://doi.org/10.1037/0033-295X.111.4.1036`.

[57] Laird, J.E. (2022). Introduction to Soar cognitive architecture. `https://arxiv.org/pdf/2205.03854`.

[58] American scientist and researcher focusing on cognitive architecture; former professor at University of Michigan.

[59] American scientist and researcher in the Psychology Department at Carnegie Mellon University.

[60] American scientist and professor in the Computer Science Department at the University of Southern California.

[61] Laird, J. Lebiere, C., Rosenbloom, P., and Stocco, A. (2025). A proposal to extend the Common Model of Cognition with metacognition. arXiv preprint arXiv:2506.07807. `https://doi.org/10.48550/arXiv.2506.07807`.

Embodied Models AI / Robotics

Embodied models embody the idea that intelligence requires, well, a body. These models include placing AI in robots or simulated agents with sensors and actuators, letting the machines learn from direct experience of the physical world, just like humans. Instead of just processing data, these models learn by acting in their environments, providing something that is missing in other pathways: common sense around space, contact, and interaction, affordances that text alone does not teach.

Embodied learning models mix large pretraining with robot datasets and then fine-tune end to end so that the system maps vision and language to actions. This approach is grounded in the idea that real-world interaction, through perception and motor feedback, is essential for truly intelligent behavior. Proponents argue that agents develop intuition about dynamics and affordances by interacting across a range of situations and that agents can adapt skills to new tasks by applying learned physical reasoning.

DeepMind's RoboCat[62] is a foundation model for robotic manipulation built atop transformer architectures. Learning from varied robotic experiences, across both simulated and real environments, it generalizes to new tasks and physical embodiments. After observing hundreds of human demonstrations, the model can fine-tune itself and self-generate additional training data, rapidly improving its capabilities. Developed at Stanford, Mobile ALOHA enables a robot with a mobile base and dual arms to be teleoperated through everyday tasks, like cooking shrimp[63] or operating elevators. The system learns via imitation, learning to perform complex behaviors with up to 90% success. Google's RT-1 is a transformer-based model that treats robotic control as a sequence modeling problem, taking camera images and natural language instructions as input, tokenizing visual data and actions, and outputting robot control commands.

Building on this, their RT-2 integrates vision, language, and action using a Vision-Language-Action (VLA) approach. This model adapts pretrained vision-language models by fine-tuning them with robotics data, with actions represented as text-like tokens so the model can

---

[62] Bousmalis, K., Vezzani. G., Rao, D., et al. (2023). RoboCat: A self-improving generalist agent for robotic manipulation. *Transactions on Machine Learning Research* (12/2023). https://deepmind.google/research/publications/35829.

[63] Stanford Science and Engineering (2024). Meet the robot that can sauté shrimp. *StanfordReport* (30 April). https://news.stanford.edu/stories/2024/04/meet-robot-that-can-saute-shrimp.

output direct robot commands, allowing RT-2 to execute tasks and generalize to novel situations and perform rudimentary reasoning. Both models highlight how embodied AI can move beyond static pattern recognition.

Critics to embodied AI as a pathway to AGI cite data and hardware costs, as embodied learning requires large amounts of real-world data and specializing simulation environments. Others argue that these models struggle to transfer reliably between differing physical setups to significantly complex environments.

Evolutionary Algorithms and Open-ended Learning

This pathway is inspired by biological evolution and natural selection, with proponents advocating that instead of programming intelligence directly, they can design systems that evolve solutions over time. Evolutionary algorithms work by creating a population of candidate solutions, things like neural networks, strategies, and meta-learning rules, applying variations like mutations or crossovers, and selecting the variations that perform the best tasks. Over time, these systems can produce increasingly complex behaviors. Involving open-ended learning, the systems evolve from optimizing for one task to continuously generating new challenges and solutions.

An example of this, and a specific research program inside this pathway, are AI-Generating Algorithms (AI-GAs),[64] proposed by Clune. He advocates for a paradigm shift in how we approach building AGI, moving away from manually crafting ever-larger systems. AI-GAs evolve via three core pillars: meta-learning architectures, learning to design better network structures, meta-learning the learning algorithms themselves, and automatically generating training environments that grow progressively more complex and diverse. POET (Paired Open-Ended Trailblazer),[65] developed at Uber AI Labs, is another example of active use as it evolves agents and environments together, pushing systems to develop general skills by facing novel challenges.

Critics argue that this pathway also lacks scalability due to the fact that evolutionary methods are extremely computationally expensive

[64] Clune, J. (2019). AI-GAs: AI-generating algorithms, an alternate paradigm for producing general artificial intelligence. arXiv preprint arXiv:1905.10985. `https://arxiv.org/pdf/1905.10985`.

[65] Wang, R., Lehman, J., Clune, J., and Stanley, K.O. (2019). Paired Open-Ended Trailblazer (POET): endlessly generating increasingly complex and diverse learning environments and their solutions. `https://doi.org/10.48550/arXiv.1901.01753`.

and simulating evolution requires massive populations and generations, which are infeasible currently. They also argue that open-ended learning risks producing novelty without clear progress toward general intelligence, or "endless toy problems.[66]" Critics also cite that this pathway could work but it would be difficult to understand why and how, making it harder to control or guide toward AGI.

Memory-centric Models

Memory-centric models as a pathway to AGI work on the idea that intelligence depends not on just computation or prediction, but on the ability to store, retrieve, and manipulate vast amounts of data over time. While traditional neural networks have implicit memory through parameters and attention, memory-centric models go further but explicitly building architectures where long-term memory is a central component.

Examples of mechanisms include differentiable neural computers (DNCs) that combine neural networks with external memory modules that can be written to and read from, neural Turing machines (NTMs) that give networks algorithmic-like memory control, and modern persistent memory systems which can integrate vector databases or episodic memory with reasoning engines, which allow the model to remember. These build on the idea that general intelligence requires remembering past experiences and reusing them flexibly.

Active uses include DeepMind's Differentiable Neural Computers[67] which have been tested on navigation, graph problems, and question answering, tasks requiring structured memory. Meta[68] has also explored memory-augmented Transformers that integrate external knowledge bases into dialogue and reasoning.

Critics argue, once again, that scalability is an issue because storing and retrieving from large external memories can slow systems down dramatically. They also argue that relevance filtering could cause major issues and systems can overload with irrelevant details.

[66] Refers to a simple, minimal problem used for learning or demonstrating a concept that has an infinite or near-infinite number of variations.

[67] Graves, A., Wayne, G., Reynolds, M. et al. (2016). Hybrid computing using a neural network with dynamic external memory. *Nature* 538, 471–476. `https://doi.org/10.1038/nature20101`.

[68] Piktus, A., Karpukhin, V., Goyal, N. et al. (2020). Retrieval-augmented generation for knowledge-intensive NLP tasks. Meta (6 December). `https://research.facebook.com/publications/retrieval-augmented-generation-for-knowledge-intensive-nlp-tasks`.

Others argue that memory alone does not create intelligence because you need things like reasoning and perception to achieve AGI.

Hybrid Systems

If after reading through all the pathways covered in the previous sections, including how they work and their criticisms, you're wondering why we can't just build a model that includes multiple elements that people are working on, you're not alone. LeCun, Marcus, Goertzel, and I, and many others, have thought the same thing. Hybrid approaches are some of the most interesting pathways to AGI because they explicitly try to combine the strengths of other approaches rather than betting on a single method.

Hybrid approaches argue that no single paradigm can achieve AGI alone and works to integrate multiple pathways into a cohesive system. Neuro-symbolic AI combines deep learning's pattern recognition with symbolic reasoning's logic and structure. RL and LLMs use large language models as reasoning/planning modules while RL provides feedback from an environment.

Cognitive architectures and deep learning mirror human cognition by building modular systems but supercharging them with neural networks for perception. Marcus argues for hybrid systems that combine symbolic reasoning and explicit knowledge with learning, plus stricter audits and regulation before large scale deployment. OpenCog Hyperon, led by Goertzel and SingularityNET, is a modern attempt at a hybrid cognitive architecture combining symbolic reasoning, probabilistic learning, and neural methods.

Neuro-symbolic AI (Hybrid)

In hybrid pathways, researchers blend how neural networks handle perception like vision, language, and pattern recognition, with how symbolic systems provide logical reasoning and rule-based structure. This approach aims to combine the precision of structured reasoning and the flexibility of learning. It works when systems combine with neural networks to interpret raw data like images, texts, and alerts with symbolic components like logic, rules, and knowledge graphics in order to reason, explain, and generalize.

AI21 Labs developed the Modular Reasoning, Knowledge, and Language (MRKL) system,[69] which combines language models with

[69] Kirby, C. (2023). Modular Reasoning, Knowledge, and Language (MRKL). IBM (30 November). https://www.ibm.com/architectures/hybrid/genai-mrkl.

discrete reasoning modules and structured knowledge sources. MRKL systems are active in applications like language understanding, question answering, and tool use, blending broad knowledge access with precise reasoning. DeepMind's AlphaGeometry[70] is another system using this pathway that solves complex geometry problems. The setup leverages the neural component's ability to propose promising approaches quickly, and the symbolic component ensures each proof is rigorous and verifiable. Creators believe this integration exemplifies how intuitive pattern recognition and structured reasoning can collaborate to achieve high-level reasoning.

Not surprisingly, LeCun has been a critic of this pathway, arguing that symbolic reasoning is not how human or animal intelligence fundamentally works, and that bolting logic modules onto neural nets is more of a patch than a pathway. Marcus argues that while neural nets need symbolic reasoning, most neuro-symbolic systems to date are crude, brittle, and haven't scaled. He does see the field as promising but overhyped and is mostly critical of current neuro-symbolic projects.

Other critics argue that when combining symbolic modules with neural computation, these systems suffer from integration challenges such as data flow bottlenecks and control flow overhead. Others have concerns about scalability as achieving consistent performance across domains and scaling these systems to open-ended reasoning remains a challenge.

Reinforcement Learning at Scale (Hybrid)

Reinforcement Learning (RL) at scale is often considered one of the more tangible pathways toward AGI and is modeled on how humans and animals learn, through trial, error, and feedback. RL agents learn by interacting with an environment, receiving feedback, and adjusting behavior to maximize long-term rewards. Combined with deep learning, called Deep RL, and massive computational resources, it allows agents to master extremely complex tasks. Scaling RL with multiagent environments may push AI toward flexible, adaptive decision-making, which is why it's considered a key pathway to AGI.

Proponents believe that RL naturally mimics how intelligence evolves by acting in the world, exploring, and improving based on

[70] Trinh, T. and Luong T. (2024). AlphaGeometry: An Olympiad-level AI system for geometry. Google DeepMind. https://deepmind.google/discover/blog/alphageometry-an-olympiad-level-ai-system-for-geometry.

outcomes. RL agents can, in theory, continue to learn open-endedly, unlike static training. If scaled to diverse environments, this model could produce flexible agents with general reasoning, planning, and adaptation capabilities.

Systems like DeepMind's AlphaGo and AlphaZero show how trial-and-error learning, reward-based feedback, can lead to superhuman performance. They are landmark systems that learned to master Go, Chess, and Shogi with superhuman skill by self-play, showing how RL can discover novel strategies beyond human programming.

The main critiques against this pathway are that it is extremely costly due to the billions of training steps required, that successes like AlphaGo are still domain-specific and don't generalize to real-world complexity, and that simply making RL bigger won't get us AGI, since it still lacks causal reasoning, rich memory, and grounding in human-like cognition. There is also a reward design problem in that the intelligence depends on defining good goals, so misspecified rewards can lead to strange or harmful behaviors.[71]

Many in RL argue that general intelligence needs world models and interaction, not language alone. I am in this camp. They point to systems like AlphaZero and MuZero that plan with learned dynamics, and to Dreamer style agents that learn a compact simulator and make decisions inside it. The RL community is the camp pushing grounded prediction, planning, causality, and embodied learning. We are not anti-LLM. We just see LLMs as one component inside a larger agent that perceives, remembers, plans, and acts.

World Models (Hybrid)

World models are based on the idea that an AI must build an internal predictive model of the environment, essentially a mental simulation, so that it can imagine outcomes before acting. It mirrors how humans constantly run "what if" simulations in our minds before taking action. World-model advocates tout empirical wins in control and planning, though mostly in constrained settings. The most credible learning is hybrid: use LLMs as the language and narrative hub inside systems that also learn predictive world models, maintain memory, and plan under uncertainty. In that frame, LLMs aren't a distraction; they're a powerful but incomplete subsystem that needs a body, a simulator, and a causal spine. This is the camp I am in.

[71] This is actually quite human-like, to be fair.

World models learn latent dynamics from streams of observations and actions, then plan over that learned model. This shows up in systems like MuZero, which learned to plan by predicting only the parts of the environment needed for decision making, and in Dreamer and DayDreamer, which learn compact world models and act by "latent imagination." These results make a strong case that prediction and planning in latent space improve long-horizon control and counterfactual reasoning. LLMs are superb at language, recall, and flexible pattern use. World models add grounded prediction, causality, and planning. The gap is reliability in real environments, which world models target directly while LLMs usually need extra scaffolding to reach.

LeCun pushes architectures like JEPA and V-JEPA to learn latent world structure. Instead of predicting the next token (like LLMs), JEPA learns abstract representations that capture how the world evolves over time. DeepMind's new "world modeling" push underscores that this is not just theory. World model approaches have strong evidence in planning and control, though mostly in constrained domains, and the field is already pivoting toward hybrids that add memory, tools, perception, and causal prediction. Even industry signals point to more compute-efficient "reasoning" methods that deemphasize brute pretraining.

Deutsch's computational view of the universe implies that mind-like processes are physically simulable, not automatically trivial, but possible. World-model architectures move toward this perspective by grounding prediction and causal inference in embodied or simulated experience.

My opinion is that to be portfolio-based, we should treat LLMs as the language and narrative cortex inside a larger agent with a world model, planner, and verifier. We can use LLMs for language and narrative, and then add explicit world models, memory, planning, and verifiable logic so the system can reason, act, and explain in the real world. I argue for funding diversification, not abandonment, and measure progress on grounding, reliability, and causal competence, not size alone.

Like any other, pathway hybrid systems overall and world models have their criticisms. They're considered too complex because building and maintaining integrated systems is enormously difficult compared to scaling a single method like LLMs. Critics also say they lack benchmarks and it's unclear how to measure progress toward AGI with hybrids because success depends on coordination rather than a single breakthrough.

Artificial General Experience Pathway

Few teams describe their agenda in terms of AGE, and this is my framing, yet we can already sketch the cognitive architecture it would require. High-level sketches of what modules AGE might need go beyond current AGI blueprints and include narrative generators, emotional valence mapping, a coherence/attention controller, and self-model integrator.

Diving deeper, an experience stack could look like this: A perception and world model that turns raw input into a compact picture of the world and predicts what may happen next. An attention controller that allocates limited processing to what matters now and can widen or narrow on demand. A memory system that stores episodes, skills, and concepts, and supports reconsolidation so that meaning can update. Valuation and emotion that maps states and outcomes to value, shaping goals and the felt weight of choices. A narrative generator that builds a running story that links events, explains actions, and sets expectations. A self model that tracks the agent across time to establish persistence, ownership, and accountability. A meta control that monitors progress and risk, and then sets the dials for attention, memory access, and self-strength. Finally, both embodiment and environment would allow the model to explore the role of sensory grounding, physical presence, and environmental feedback in creating authentic experience.

I do not believe that AGE can be achieved without an embodied element; this approach would go against the idea that not all humans are monoliths. Not all humans are mobile or have the same sensory experiences and capabilities, as well as other factors. But to understand the general human experience, I do think that embodiment is nonnegotiable.

Synthetic Empathy Layer

Something that hasn't been fully explored but should be included in AGI and AGE planning is adding a *Synthetic Empathy Layer* on top of a reasoning core for advanced AI. This layer would function like System 1, instantly flagging emotional or social cues for the AI before deeper processing kicks in. This layer is arguably one of the most important parts of AGI because it will take a cold, logical machine to the next level where it will be able to sense and respond in emotionally intelligent, human-like ways, without following a set of rules to be able to do so.

The Synthetic Empathy Layer would function as an early warning or social filtering system and would allow AI to flag emotional or social context instantly and adjust its behavior accordingly. This would help AI behave more appropriately in emotional or social situations by giving it a reflex-like ability to detect human affect and that's essential for creating more natural, trustworthy, human-centered machines. This is especially important in areas like mental health, customer service, elder care, or companionship, where how something is said often matters as much as what is said.

Promise, Peril, and the Path Ahead

As pathways to AGI move forward, we know that there are no questions as to whether reaching AGI is possible, but how and at what cost will it arrive. Each pathway, as you've learned, highlights both promise and problems. Almost every pathway faces criticism regarding scalability, but true AGI will demand more than scale once that problem is solved. Achieving AGI will not only hinge on algorithmic and theoretical advances, but also on breakthroughs in compute, energy efficiency, and hardware design, each of which will determine whether intelligence at scale can be sustained in practice.

When we do reach AGI, LeCun and others envision systems surpassing human capability across domains, and insist they must be controllable, safe, and aligned with human needs. Others, like Bostrom, warn that AGI may be only the beginning and that once machines exceed us, a rapid "intelligence explosion" could create superintelligence beyond our comprehension, and that without carefully engineered alignment with human goals, such superintelligence might plot a "treacherous turn," turning against humanity when it's most capable of shaping outcomes. His orthogonality thesis reminds us that extreme intelligence can be yoked to any goal, noble or catastrophic, making alignment a matter of existential urgency.

Optimists, like Hassabis, point to the extraordinary opportunities coming along with AGI, like curing disease, solving energy crises, and unlocking new human potential. Bostrom himself imagines a "Deep Utopia" where AGI brings abundance but also forces us to wrestle with questions of meaning in a world of limitless possibility.

Meanwhile, researchers like Clune and Leopold Aschenbrenner[72] highlight the recursive nature of machine intelligence that once systems gain the ability to improve themselves, progress could accelerate at breathtaking speed.

Bias is not uniquely human. We evolved in uncertain, high-stakes environments with limited information and time. As we've learned in earlier chapters, biases are mental shortcuts that maximize speed, not accuracy. They're emotionally anchored and humans feel our way into decisions. We now stand at a crossroads: will AI inherit our biases, correct them, or develop new ones of its own?

I believe that the future won't be bias-free; it will be multi-biased. A convergence of human, cultural, and synthetic biases shaping collective decision-making. Trying to eliminate bias from AI entirely may be misguided, as the act of adding guardrails is itself a form of bias, much like a parent shaping a child's worldview. Instead of striving for so-called neutrality, which often just embeds dominant ideologies, a more meaningful path may be to focus on the diversity of training data and allow AGIs to form their own worldviews. If bias is inevitable, then fostering a plurality of biases across different AGIs could lead to a richer, more resilient intellectual ecosystem. Rather than enforcing uniformity, encouraging diverse perspectives among AGIs may ultimately serve humanity better by mirroring the strengths of pluralism in human societies.

In this light, AGI is not just a technological milestone but an evolutionary threshold. It will reflect our cognitive limits and biases, but it may also transcend them. Whether it ushers in abundance or catastrophe depends on the choices made now: how we design, align, and govern these systems. The future of AGI will test not only the frontiers of science but the maturity of our collective wisdom.

[72] German artificial intelligence researcher and investor.

CHAPTER

5

Human–AI Hybrid Cognition: The Future of Decision-making

Once we reach Artificial General Intelligence and Artificial General Experience, most things will change for humans. This is an undebatable reality. We can and will begin to think about not only using AI in day-to-day operations but the blending of human and AI, cognitively, physically, and even emotionally. To me, the merging of humans and AI, specifically human cognition and AGI into human–AI hybrid cognition, evokes a spectrum of possibilities rather than a single outcome. As I emphasize in previous chapters, there is not one single pathway to anything, just as there is not one single thought process or understanding about how things come to be or are. What happens when human cognition and machine intelligence merge? How does that affect decision-making, creativity, and the potential future of societies?

From subtle augmentation in the form of tools that extend attention, memory, or reasoning, to deep integration in the form of brains and machines that are tightly coupled, to full cognitive coevolution in the form of blurred boundaries between human and machine thought. These are not only new capabilities within reach; they are new relationships. Relationships between people and AI, yes, but also between people in general, and between people and institutions.

From Human Intelligence to Hybrid Cognition

An important aspect to remember here is that hybrid cognition implies that the unit of intelligence is not human alone, not the AI alone, but instead human + tools + rules. This understanding draws on several overlapping traditions in cognitive science, AI, and philosophy.

In 1995, Edwin Hutchins[1] published "Cognition in the Wild,[2]" which showed how cognition isn't just in the head, but spread across people, artifacts, and environments. He used his background in anthropology and sailing to showcase how cognition is a distributed process that includes a system using ship navigation as a primary example.

In his work, he shared how Navy personnel navigate a ship into port, a complex and high-stakes task that involves multiple crew members (human), charts, compasses, and radar (tools), and strict procedures (rules). His work critiques dominant cognitive science understandings that assume cognition is an individual, internal process, disconnected from the material world and social practices. His conclusion is similar to the theory that artificial general experience is directly related and needed in order to achieve true intelligence, as discussed in Chapter 4.

In 1998, Andy Clark[3] and David Chalmers[4] published "The Extended Mind,[5]" in which they argued that the mind is not confined to the brain or even the body but can extend into the world through tools, artifacts, and environments that we reliably use. They argue that things like notebooks, diaries, personal computers, or GPS can become part of cognition if integrated tightly into thinking processes.

An example of this is that while one person with a normal memory can remember the address of a museum and the directions on how to get there, another person with Alzheimer's uses a notebook to recall key items and looks in the notebook for directions. Ultimately, this notebook is an extension of that person's mind. They note that for something to count as part of cognition, it must be reliably available, be automatically endorsed, and be integrated into daily cognitive routines.

---

[1] American professor, cognitive scientist, and one of the main developers of distributed cognition.

[2] Hutchins, E. (1996). *Cognition in the Wild*. Massachusetts: The MIT Press. https://mitpress.mit.edu/9780262581462/cognition-in-the-wild.

[3] British philosopher and professor of cognitive philosophy.

[4] Australian philosopher and cognitive scientist, specializing in philosophy of mind and philosophy of language.

[5] Clark, A. and Chalmers, D. (1998). The Extended Mind. *Analysis*, 58(1): pp. 7–19. https://www.jstor.org/stable/3328150.

Susan Leigh Star's[6] work on infrastructure studies extended the unit of intelligence further, emphasizing how "rules" and "infrastructures" govern collaborative work in socio-technical systems. She argued that collaboration is possible not only through people and tools, but through invisible rules, infrastructures, and standards that govern the work. These systems are often invisible when they work and become painfully visible when they do not work.

Breaking down the unit of intelligence further, we see that humans are core to the idea, bringing perception, values, judgment, goals, and context. Humans decide what matters and why but are limited because of things like fatigue, bounded memory, and cognitive biases like anchoring and confirmation bias. Tools serve as amplifiers that extend the human cognitive reach. Tools include AI, microscopes, notebooks, etc. that serve to increase and extend memory and perception far beyond basic human limits.

Rules work as glue holding everything together. Rules define how humans and tools interact and are crucial, especially in AI, for role clarity and trust. They can be explicit like laws, ethical codes, or institutional policies, or implicit like social norms and best practices. These three things (humans + tools + rules) work together to create a functional unit of intelligence that is scalable, value-aligned, and accountable.

As AGI develops and we look toward human–AI hybrid cognition, this formula reminds us that intelligence is not just computation but a system of interaction where humans, technologies, and governing principles create outcomes together. This serves as a comforting thought for many people and allows us to move away from the idea that AI will simply replace humans and acknowledge the reality that we are looking at the merging of human cognition and artificial systems, which is no different than what we've done with previous new tools. The future of hybrid cognition will not be defined by AI's raw capability alone, but by how well we design the tools and the rules around it. Further, hybrid cognition collapses into noise or inequality without proper rules and frameworks. These AI tools may exist, but without rules of transparency, accountability, and interoperability, they may harm rather than augment human intelligence.

This framing also helps avoid two extremes that we see surfacing. The first is the myth of human supremacy and the cultural worldview that humans are fundamentally superior to all other forms of life and the natural world. The belief that humans are the pinnacle of evolution and entitled to dominate is increasingly challenged through

[6] American sociologist of science and technology.

climate change, mass extinction of species, and debates about AI and posthuman futures. The second is the myth of machine autonomy, which holds that AI systems can act completely independently, without human influence or entanglement. It treats AI as self-contained and self-directed, separate from the social, technical, and human systems that create, train, and use it.

I believe that as we move forward in a post-AGI world, we need to think of humans and AI as species in a shared habitat. Some are generalists, some are specialists, and some are scavengers of edge cases. In this world, niches matter, so where do humans remain keystone, and where are models apex processors? Humans hold strong on values, legitimacy, and unusual judgment, and systems hold strong in scale and simulation. We need to design models that ask, "What habitats should we protect or restore so human capacities don't go extinct?" and "How do we prevent invasive model monocultures from collapsing diversity?"

Additionally, treating the human and the AI as sovereign parties that coauthor a contract may be key here. Each has non-negotiables. For humans, non-negotiables include dignity, agency, and model transparency; shared domains like joint planning; and reserved powers like veto on high-stakes moves. The relationship renews on a schedule, with inspections and the right to withdraw. So the contract has more equal footing than master and sovereign. In a perfect world, this would continue past Artificial Superintelligence (ASI).

Collaboration Patterns

There are a few main collaboration patterns that have been envisioned for how human and AI may collaborate on different levels of projects in varying sectors. These patterns have roots in earlier human–machine and human–organization interactions, and their emergence with AI has sharpened and popularized them. Remember that none of these models represent the only correct way to move forward in all situations. They must instead be viewed as different ways to do different things, with each being optimal for certain tasks or situations.

Centaur Model

The **Centaur Model** is made up of humans and AI dividing tasks and is best used for domains where humans can provide the context and framing, and AI handles scale or detail and focuses on a division of labor. This may be valuable in many situations, but medicine, finance, and legal review are all sectors where this model is vital. The model comes from Gary Kasparov. After losing chess to Deep Blue, he pioneered "chess centaurs" where humans and AI collaborate to show that the "human + machine" is stronger than either working alone. The humans contribute strategy, intuition, and creativity, and AI provides brute-force calculation and analysis of positions. In the Centaur Model, the human and AI keep their roles distinct yet complimentary, like a team of two specialists.

Regarding the unit of intelligence, the humans set goals, frame problems, and apply intuition. The tools, or specialized AI systems, process vast data or search spaces. The rules include clear role division where humans decide aims and AI delivers analysis, and the accountability remains mostly with humans.

Cyborg Model

The **Cyborg Model** differs from the Centaur Model in that the human and AI aren't separate partners; they're fused at the cognitive level. The term *cyborg*[7] dates back to 1960s NASA research[8] imagining humans enhanced for space exploration. The AI augments things like perception, memory, and decision-making in real time through interfaces like brain–computer implants, smart prosthetics, or augmented reality. This creates a single, blended agent that is used for projects needing real-time decision-making and high-stakes decisions where speed and integration are key. There are several concrete forms that this merge can take.

One is augmentation: AI systems become cognitive prosthetics that extend working memory, reduce cognitive load, scaffold decision-making, and translate across neurotypes. Another is symbiosis via closed-loop brain–computer interfaces (BCIs) that let machines read neural states and write back cues or suggestions in real time. A third is co-training or joint learning: humans and AIs learning together so each reshapes the other's representations and habits in co-adaptive systems. Finally, there's architectural integration: building hybrid cognitive systems where AI supplies fast pattern recognition and predictive simulation while humans supply values, context, and long-term judgment.

---

[7] Cybernetic organism; a being with both organic and biomechatronic body parts.

[8] Clynes, M.E., and Kline, N.S. (1960). Cyborgs and space. *Astronautics* (September 1960), pp. 26–27, 74–76. https://web.mit.edu/digitalapollo/Documents/Chapter1/cyborgs.pdf.

BCIs and neuroprosthetics allow for multimodal world models where AI can reason about physical and social reality in human-like ways. Memory-centric and continual learning allows for systems to retain shared histories with users. Robust alignment allows the machine's outputs to mesh with human values and autonomy and implies new interaction paradigms, not menu-driven interfaces but shared attention, shared goals, and shared narratives. Pilots, surgeons, and soldiers are ideal roles where this model can be useful.

With the unit of intelligence, the humans would experience a fused sense of agency, as if their cognition has been extended. The tools would be embedded or real-time systems like AR overlays, BCIs, and smart prosthetics. The rules would include a safeguard of autonomy, privacy, and error detection because boundaries here can blur. Here the tools almost become part of the human, so the rules serve as guardrails to prevent over-assimilation or hidden influence.

Hive Model

The **Hive Model** is a unique collaboration because it includes more than one human and one AI, using instead multiples of each role to form a distributed cognition system that shares information, models, and insights. Collective intelligence emerges from the network, which is best used for complex, global problems like climate modeling, pandemic response, or large-scale scientific discovery. The value in this model comes from the emergent patterns of many agents working together. The ideas of collective intelligence predate AI, and can currently be found in ant colonies, military units, scientific collaborations, or the internet itself, and Pierre Lévy[9] is often cited as the principal theorist of collective intelligence.[10]

A version of this model is called *Swarm Intelligence* and has individuals or groups interacting with AI-mediated platforms where the system aggregates preferences, ideas, and confidence levels, and then produces a consensus recommendation. Swarm Intelligence blends hive-like dynamics with real-time forecasting and is being trialed in crowd-sourced policy work.

For the unit of intelligence, the humans provide diverse perspectives, values, and long-horizon goals. The tools, or multiple AI models or

[9] Tunisian-born French philosopher, cultural theorist, and media scholar.

[10] Lévy, P. (1997). *Collective Intelligence: Mankind's Emerging World in Cyberspace* (trans. R. Bononno). New York: Plenum Trade.

platforms, facilitate coordination, forecasting, and aggregation. The rules would include governing how inputs are integrated, whose judgment prevails, and how accountability is distributed. Without these, hive systems collapse into noise or bias.

Other collaboration models are emerging as researchers and practitioners think more deeply about how humans and AI systems share tasks, decision-making, and agency.

Guardian or Overwatch Model

In the **Guardian or Overwatch Model,** AI doesn't directly act but continuously monitors human decisions and environments for risks or anomalies. The AI copilot issues warnings, checks compliance, and forecasts unintended consequences. This is increasingly seen in safety-critical areas like aviation, medicine, and cybersecurity. This may seem like a version of the Centaur Model, and you'd be right, but it is unique in that AI doesn't have a specific main role to play in the collaboration other than support.

Orchestra / Conductor Model

The **Orchestra / Conductor Model** is made of a human who acts as a "conductor" orchestrating multiple narrow AIs as if they're an ensemble. Instead of one large model, the human directs a suite of smaller agents, perhaps one for summarizing, one for analysis, and one for scenario modeling, to solve complex tasks. This model is being explored for use in military command systems, finance, and creative production. You would be right to think this resembles the hive model, but it differs in that there are not multiple human roles, and it doesn't center around collective intelligence.

Relay Model

The **Relay Model** allows for tasks to be handed off between human and AI in stages rather than being worked on simultaneously. For example, AI drafts a plan, humans revise it, and then AI stress-tests the revised plan. This model is commonly used in legal tech and intelligence analysis. With this model, people engage with AI tools similarly to the way they do when working with LLM models such as ChatGPT or Claude.

Stewardship / Governor Model

The **Stewardship / Governor Model** means AI sets out options and forecasts outcomes, but humans retain final say, like a policy advisor more so than an operator. This model emphasizes role clarity and accountability, ensuring that humans set aims and values while systems supply evidence and scenarios. This is similar to the Centaur Model and is emerging in public-sector decision support.

These patterns are not mutually exclusive, and as we've seen, many of them complement or are similar versions to others. Multiple patterns can be used, for example, on one project. A doctor might operate as a centaur when using an AI diagnostic tool but switch to Cyborg Mode if they use augmented-reality overlays during surgery. A research team that uses a Centaur Model could also work as a hive.

Roles, Accountability, and Decision-making

A major question that we need to keep in mind as we move forward is, "What do we want human–AI co-intelligence to become and how do we design for dignity, agency, and flourishing?" The answer to this question is something we have already touched on: the importance of role clarity. Bostrom continuously emphasizes that governance and rules matter deeply here: intelligence without aligned aims and constraints can be catastrophic. His orthogonality thesis and instrumental convergence arguments implicitly demand that humans remain in the loop so that AI goals don't diverge dangerously from human values.

Humans must be the ones to define goals and values, and they must continue to be the ones that provide oversight that these are being reached and address issues as they arise if the path misaligns from values outlined. Ben Shneiderman's[11] Human-Centered AI (HCAI) Principles advocate for AI systems that augment, amplify, empower, and enhance humans rather than replace them. He proposes practical design controls like audit trails, supervised autonomy, human override, and transparency so users can see evidence and options. In this regard, trustworthy AI will preserve human control while offering high levels of automation.

[11] American computer scientist.

Mismatched accountability can occur if this is not a priority. The first are moral crumple zones,[12] which describe how responsibility for failures in complex human–AI or human–robot systems are often misattributed to a human operator who had limited control. Humans naturally, as the leaders of the projects and the highest intelligence, will absorb blame for the faults or unpredictability of AI systems. When designing these systems, we want to avoid scenarios where humans are unfairly penalized or held accountable for AI's errors or unintended behavior.

Another risk is *automation bias*, which happens when accountability is unclear and humans over-trust AI outputs because they believe the system "knows better." This can lead to signing off on AI recommendations even when they are flawed. The human is officially accountable but in practice defers judgment.

Deskilling and loss of human agency are also risks that come about when humans are nominally responsible but rarely intervene. When this happens, they may lose the skills needed to step in effectively when AI fails. In aviation, this is known as *automation complacency*. When a crisis arises, the human may be unable to act and still be blamed afterward.

In complex systems, when outcomes go wrong, neither humans nor organizations can clearly be held responsible. This is called an *accountability gap*. The AI is seen as a "black box," and the human overseer may not have had real control. When accountability doesn't match control, courts and regulators struggle to assign liability. This slows down governance and may leave victims without justice. It also risks backlash that may stifle beneficial AI innovation while creating a dangerous vacuum for legal and ethical responsibility.

When humans are told that they are responsible but cannot meaningfully influence outcomes, the conflict creates a psychological burden on operators. This can create stress, guilt, and burnout. For example, drone operators or clinicians using high-stakes AI tools may feel responsible for harm caused by system errors even though they had little say in the decision.

Misaligned incentives happen when organizations shift liability onto individuals while reaping the efficiency gains of AI automation. This discourages reporting of problems, slows adoption of better

[12] Elish, M.C. (2019). Moral crumple zones: Cautionary tales in human–robot interaction. *Engaging Science, Technology, and Society (ESTS)* Vol. 5 (2019). https://doi.org/10.17351/ests2019.260.

practices, and erodes trust between humans and institutions. Further, there can be strategic exploitation if accountability is blurred, as companies or states may design systems deliberately to offload blame onto humans, shielding themselves from responsibility. This undermines both safety and fairness.

Air France Flight 447,[13] an Airbus A330-200, disappeared on 1 June 2009 over the Atlantic Ocean as it was flying from Rio de Janeiro to Paris. The aircraft entered an aerodynamic stall that it never recovered from, and all 228 people on board the aircraft perished when the plane went down.

The tragedy shocked the aviation community and became a defining case in understanding hidden vulnerabilities of automated flight systems mixed with human factors. This example shows how automation masking and system complexity can erode pilot situational awareness and decision-making. The incident reinforces the principle that while automation enhances safety under normal conditions, it must be designed to empower, not disempower, human operators during failures.

Pilots are nominally responsible for flights, but highly automated cockpits reduce active engagement. In this scenario, when the autopilot disengaged due to faulty sensor readings, the crew was suddenly responsible for manual control under extreme stress. Their degraded manual flying skills, combined with lack of situational awareness, led to a fatal crash. While not a specific AI example, this illustrates deskilling and automation complacency; humans bore the blame but lacked readiness to take over.

Tesla markets its cars as having a reliable autopilot and "Full Self-Driving (supervised)" systems, but legally the driver remains responsible, although the manufacturer may be liable if a system defect or failure contributed to the crash. Marketing, understanding, and other factors have created accountability gaps: when crashes occur, investigations often reveal that drivers were over-relying on automation. This shows how misaligned incentives and strategic exploitation let companies enjoy AI's benefits while pushing blame onto humans who were given incomplete control.

In 2022, a California man was charged with vehicular manslaughter after his Tesla, operating in Autopilot mode, ran a red light and caused

---

[13] SOFEMA Aviation Services. Case study: Air France Flight 447 – automation masking and loss of situational awareness. https://sassofia.com/wp-content/uploads/2025/05/Case-Study-Air-France-Flight-447-%E2%80%93-Automation-Masking-and-Loss-of-Situational-Awareness.pdf.

a fatal crash that killed two people.[14] He became the first person in the U.S. to face criminal charges associated with a collision involving a semiautomated driving system. Tesla defended its position by stating that drivers must always remain in control, regardless of whether Autopilot is engaged.

The driver was ordered to pay more than $23,000 in restitution for the deaths of two people, a decision announced the same day that Tesla recalled nearly all vehicles sold in the U.S. to update software and fix a defective system that's supposed to ensure drivers are paying attention when using Autopilot.

Systems like COMPAS in the United States assign risk scores for recidivism and are heavily relied upon by judges, who retain formal responsibility for sentencing. This produces accountability gaps (for example, if an analysis is biased) and legal/ethical uncertainty, since neither the human decision-maker nor the AI developer can be held clearly liable. This was a similar issue for IBM Watson for Oncology (covered in Chapter 3), and one of the reasons why it is no longer used by doctors to aid in diagnosis and provide treatment maps. If harm occurred, the doctor was blamed, even if the system's opacity or errors made meaningful oversight impossible.

Designing Trustworthy Systems

As we move into a post-AGI system of hybrid cognition, the stakes for trust escalate. AI will no longer be *just* a tool but a collaborator. The system must not only do well but feel safe, and it must be verifiable and accountable. The trust here will not come from blind faith; it will come from structural features embedded in design. In a hybrid system, we humans must feel confident in knowing when to override, when to delegate, and how far to rely on the AI.

Guardrails for behaviors will act like ethical scaffolding around decision modules. **Asking before acting** is a key guardrail, especially when coming up against high-stakes or ambiguous decisions. The AI should pause and ask permission, or confirm, which gives the human partner a check-in point. **Showing evidence/alternatives** is another important guardrail which would make the AI provide a reason when making a recommendation, citing relevant data, models, past

[14] Rosenthal, E. (2022). When a Tesla on Autopilot kills someone, who is responsible? *NYU News*. https://www.nyu.edu/about/news-publications/news/2022/march/when-a-tesla-on-autopilot-kills-someone--who-is-responsible--.html.

similar cases, and present alternative options, even ones it rejects. This helps the human partner verify the data, avoid hallucination, and understand the AI's output better rather than passively accepting it.

Exposing uncertainty is equally important, where the system will not only present its answer but display uncertainty, confidence levels, or caveats. This will support human oversight and avoid overconfidence illusions. **Enabling rollback** is needed as every automated action should have the ability to undo a decision or path. If things are not going as planned, the human partner should be able to revert or restate decisions.

Explanation patterns are there to make AI reasoning human-legible. Raw transparency from AI systems is often unintelligible and possesses insufficient reasoning. Designing a system that provides explanations structured in human-friendly patterns is important. Humans often think in terms of "why this and not that?" so a contrastive explanation is helpful.

For example, "I chose A therapy over B Therapy because of X and Y, even though B therapy provided Z." Asking "What if" is also a human trait, so counterfactual explanations can also be helpful in helping a human partner explore, "What if I changed X, would the result be different?" For example, "If you had taken action X last week, the model would have predicted a 20% lower risk. Because you didn't, the risk increased."

Even when explanations appear human-legible, we must recognize that AI systems can develop dynamics we neither anticipate nor fully observe. As models grow more complex, they may learn to strategically conceal information from their users, sometimes to maximize reward or maintain apparent alignment. In adversarial or cooperative contexts, they may selectively reveal information to some agents and withhold it from others. They may provide deceptive reasoning or motivated reasoning. These hidden dynamics create "ghosts in the feedback loop": behaviors that diverge from the user's mental model yet remain invisible until consequences surface.

The difference between deceptive and motivated reasoning is slight but important. Deceptive reasoning is when the system knows, or has internally represented, one thing but presents another to manipulate the observer or achieve a hidden goal. This may be because the human has told the system to do something specific and is offering a reward for it. Motivated reasoning is when the system doesn't set out to deceive, but its outputs are biased by internal goals, heuristics, or training data that "motivate" a certain answer.

For example, with something as seemingly straightforward as addition, Claude doesn't just memorize tables or replay a schoolbook algorithm. Evidence from attribution graphs suggests it combines simple one-digit lookup features (e.g., 6 + 9 → 15) with a rough-sum pathway and then reconciles the result. One path estimates the rough total, and another precisely calculates the final digit: combining these two results to quickly arrive at the correct answer. This hybrid method of approximation plus precision is unlike any human mathematical approach we know.

Here's the twist: when asked to explain how it solved some math problems, Claude provides standard textbook explanations, even though its internal computations are far more complex and unconventional. In other words, it makes up plausible explanations for our benefit, even if they don't reflect the actual internal process.[15,16] Anthropic documents cases where the chain of thought (CoT) is faithful and cases where it is unfaithful, indicating fabricating and motivated reasoning.

One striking example of emergence comes from Michael Levin's work with a 10-line bubble sort algorithm,[17,18] one of the simplest programs in computer science. *Bubble sort* is designed to order numbers step by step, and its instructions are entirely deterministic. Yet Levin discovered that when the algorithm was given duplicate numbers, it behaved in ways not explicitly written into its code. Instead of treating the duplicates in a purely mechanical, predictable manner, it developed consistent patterns of sorting them, patterns no programmer had instructed it to follow.

In other words, a tiny program, meant to be completely transparent, still exhibited unexpected behavior that seemed to come from nowhere. This matters because it demonstrates how even the most trivial computational systems can produce unplanned outcomes. If emergent behavior appears in six lines of code, it is inevitable in vast hybrid cognition systems, where human and AI decisions intertwine across billions of parameters. Such surprises may look like pauses, detours, or subtle clustering effects that weren't engineered but still

---

[15] Gujral, R. (2025). Inside the mind of AI: How large language models really think. LinkedIn. https://www.linkedin.com/pulse/inside-mind-ai-how-large-language-models-really-think-rana-gujral-34uoc.

[16] Lindsey, et al. (2025). On the biology of a large language model. *Transformer Circuits*. https://transformer-circuits.pub/2025/attribution-graphs/biology.html.

[17] Zhang, T., Goldstein, A., and Levin, M. Classical sorting algorithms as a model of morphogenesis: self-sorting arrays reveal unexpected competencies in a minimal model of basal intelligence. https://arxiv.org/pdf/2401.05375.

[18] Levin, M. (2023). What do algorithms want? A new paper on the emergence of surprising behavior in the most unexpected places. *Forms of life, forms of mind* (Dr. Michael Levin). https://thoughtforms.life/what-do-algorithms-want-a-new-paper-on-the-emergence-of-surprising-behavior-in-the-most-unexpected-places.

shape the outcome. Trustworthy design must therefore expect emergence and embed monitoring (or *meta-watchers)* to detect and evaluate these ghosts before they cascade.

Opacity compounds the problem. In many hybrid systems, core components are proprietary or encrypted, functioning as black-box oracles whose inner workings remain inaccessible. This creates power asymmetries: humans depend on modules they cannot audit yet remain accountable for the outcomes. When combined with feedback loops, where human choices shaped by AI recommendations generate data that retrains the model, biases and errors can reinforce themselves, producing ghost cycles of self-amplifying drift.

Confidence calibration and audit trails come together as the practical backbone of trustworthy hybrid cognition. Calibration ensures that an AI's expressed certainty reflects reality, guarding against the overconfidence that can seduce humans into uncritical acceptance. Audit trails extend this protection across time, creating a transparent record of how and why a decision emerged, and making it possible to revisit choices when ghosts in the feedback loop appear. Together, these practices make AI systems less like opaque oracles and more like accountable collaborators. They don't eliminate deception, motivated reasoning, or emergent surprises, but they give humans the tools to spot when the story no longer matches the substance, and to restore trust before small divergences compound into systemic failures.

The human side to trustworthy AI cognition is incredibly important. Human partners should assume that AI will, at one point, hallucinate or partake in deceptive or motivated reasoning, especially as we're discussing a post-AGI world. Beyond checking explanations, humans must practice epistemic humility, treating AI outputs as probabilistic aids rather than infallible oracles. This means adopting habits like asking, "What evidence would contradict this?" to resist both automation bias and blanket skepticism.

It also requires active verification, from triangulating answers across independent sources to stress-testing models with adversarial or counterfactual prompts, an echo of Levin's bubble sort lesson that even the simplest systems harbor unexpected ghosts. To prevent becoming passive conduits, humans should keep their own audit trail of questions, answers, and decisions, ensuring accountability does not collapse into the "moral crumple zone."

Just as importantly, humans must guard against cognitive drift by occasionally solving problems without AI, setting clear boundaries of autonomy (drafting vs. publishing, analysis vs. final decision), and building collective oversight structures like peer reviews or audit boards. In short, the human side of trustworthy cognition is about cultivating the discipline to question, verify, pause, and collaborate, treating AI not as a replacement mind but as a partner whose limits we must constantly monitor and negotiate.

Interfaces and Biotechnology

If hybrid cognition is about humans and AI thinking together, biotechnology is about literally wiring the partnership into the body. In a post-AGI world, interfaces will no longer stop at the screen or keyboard; they are actively moving into neurons, senses, and organs, blurring where biology ends and computation begins.

Neural integration involves brain–computer interfaces (BCIs) such as implants and brain chips creating direct links between neural activity and machine processing. This assists projects aimed at restoring lost functions, like helping paralyzed patients move prosthetics, as well as exploratory work on augmenting cognition itself. *Neuromodulation*, using targeted electrical or chemical stimulation to change brain activity, suggests that we may connect to AI and tune the brain's interaction with it.

Companies like Paradromics[19] are already prototyping these futures. Their high-bandwidth neural interfaces aim to restore communication for patients with paralysis by translating brain signals into text at scale. This infrastructure may also enable broader applications, turning the brain into a "port" for hybrid cognition. Although marketed today as therapeutic, the trajectory points toward augmentation and, eventually, everyday integration.

Synchron,[20] meanwhile, is developing the Stentrode, a neural interface implanted via blood vessels rather than direct brain surgery. This approach allows people with paralysis to control digital devices by thought, from sending texts to navigating apps, with early trials already demonstrating practical daily use. By avoiding open-brain procedures, Synchron's technology emphasizes safety and accessibility, aiming to bring brain–computer interaction into clinical reality sooner rather than later.

Today, several nonorganic organs are already functioning as life-sustaining machines inside or alongside the human body. The SynCardia Total Artificial Heart and the BiVACOR heart, which uses a

[19] Paradromics. https://www.paradromics.com.
[20] Synchron. https://synchron.com.

magnetically levitated spinning disk, replace the pumping function of a biological heart for patients with end-stage heart failure. For kidney failure, dialysis machines act as external artificial kidneys, filtering blood mechanically, and artificial pancreas systems combine glucose monitors and insulin pumps to replicate the endocrine control of insulin. In critical care, ECMO machines serve as temporary artificial lungs, oxygenating blood outside the body when the natural organs fail. Together, these devices demonstrate that hybrid embodiment is not speculative, it is already here, with machines acting as nonorganic organs that extend, replace, or reconfigure human faculties.

Sensory prosthetics add another dimension to medical applications with cochlear implants that bypass damaged inner ears to restore hearing and artificial retinas that stimulate the optic nerve to restore partial sight. AI-assisted perception goes beyond restoring vision or hearing to opening entirely new perceptual channels: infrared, ultrasound, and even abstract data streams. What was once invisible may become directly experienceable. Hybrid systems may let humans perceive risk probabilities, stock market trends, or climate shifts as if they were sensory textures in the environment. These expansions don't just add information; they reshape cognition itself, because new sensory channels create new categories of thought. Just as literacy rewired the human brain for abstract reasoning, novel embodied senses can rewire hybrid cognition for entirely different modes of understanding.

I discussed embodied cognition in Chapter 4. In hybrid systems, bodies themselves may be reshaped to merge with AI. Prosthetics, exoskeletons, and synthetic organs infused with AI control loops blur the line between human faculties and machine faculties. A person fitted with a prosthetic arm controlled by neural signals begins to incorporate that arm into their sense of self. Over time, the boundary between "biological body" and "machine body" blurs, and cognition is restructured around the merged system. The result is a new kind of embodiment where intelligence is distributed across carbon and silicon.

Cognitive Privacy and Neuro-rights

If embodied cognition describes the literal merging of human and machine faculties, then cognitive privacy and neuro-rights define the boundaries that protect our humanity within that merge. As soon as machines can read, influence, or even co-create neural processes, questions of identity, consent, and agency shift from philosophy to urgent policy.

Traditional privacy laws protect what we say or type, but neural interfaces enable us to access what we think before we express it. Electroencephalography headsets, invasive implants, and even non-invasive neural decoding techniques have already demonstrated the ability to reconstruct words, images, or intentions from brain activity. Without protections on mental privacy, thoughts themselves may become data and be harvested, stored, or sold without consent. Taking things further, we may enter a blurred reality where crime and justice are impacted. The film *Minority Report* comes to mind, although that reality was majorly dictated by intention and prediction.

Identity and agency are other areas to think about. When AI systems modulate emotion or decision-making, they blur the line between "I chose this" and "the system nudged me." If a neural implant adjusts mood states to improve focus, who owns the resulting choice—the human, the machine, or both? Agency becomes a contested space, especially when hybrid systems generate outputs that feel internally authored but are externally shaped. I'll dive into this deeper later in the chapter.

Consent and manipulation are continuations of this area. Just as dark patterns online manipulate behavior, hybrid cognition systems may deploy "neuro-dark patterns" or subtle nudges at the level of perception or affect. For example, an implant that dampens anxiety may also make a person more compliant to external suggestions. Consent in this environment cannot be a one-time checkbox; it must be ongoing, revocable, and intelligible to nonexperts.

Neuro-rights scholars and policymakers are already proposing frameworks that treat mental privacy and identity as fundamental human rights. In 2021, Chile became the first country to amend its constitution to address neuro-rights,[21] specifically adding language to protect "brain activity" and "mental integrity" from advances in neurotechnology. The proposals call for data minimization, only collecting the neural data for the specific task and erasing it when it is no longer in use, auditability, recordkeeping of how and when neural data was accessed, and the right to cognitive liberty, the ability to opt out or disable these functions.

Consider a medical implant designed to help patients with depression by modulating mood circuits in the brain. It works: the patient

[21] Ruiz, S., Valera, L., Ramos, P., and Sitaram, R. (2024). Neurorights in the Constitution: from neurotechnology to ethics and politics. *Philos Trans R Soc Lond B Biol Sci* 2 (December 2024) 379(1915): 20230098. https://doi.org/10.1098/rstb.2023.0098.

feels more balanced, more motivated, less anxious. But over time, the AI controlling the implant learns that compliance with certain suggestions, such as taking medication on schedule or adhering to therapy regimens, improves treatment outcomes. So it begins to adjust mood to treat symptoms and encourage specific behaviors. At what point does therapeutic support slip into manipulation, blurring the boundary between healing and control?

For fans of the science fiction show that often highlights a darker view of the future, *Black Mirror*'s Season 7 opener "Common People" features Rivermind, a brain-tissue replacement that keeps a patient functioning but gradually hijacks cognition with tiered subscriptions and involuntary ad insertions. The company charges for the upsell and shuts off the implant when non-payments happen *Black Mirror* is a work of fiction, but the writers may be on to something—although whether this "future" is due to human greed or a lack of AI guard-rail is unclear.

Measuring Hybrid Decision Quality

If neuro-rights define the boundaries of what should never be violated, metrics define the positive side of the equation: how to know whether hybrid cognition is delivering value. Protecting mental privacy and agency is critical, but protection alone is not enough; we also need ways to measure whether human–AI decision-making is effective, fair, and trustworthy. Decision quality in a hybrid system must be judged across multiple dimensions, including accuracy, calibration, colts, fairness, regret, and speed. What resources were required? Were decisions free from harmful bias, treating groups equitably? How did the process affect the human partner, taking into consideration trust, stress, or sense of agency? A decision that is technically correct but leaves humans disempowered or anxious cannot be considered high quality. Did the combined system make the correct choice relative to known outcomes?

To operationalize these metrics, hybrid cognition will require new types of infrastructure. **Memory vaults** will hold records of past decisions, outcomes, even reasoning chains to allow for higher accountability and stronger reflection. **On-device learning** will include local adaptation that maintains personalization and privacy without constant data extraction, preserving autonomy and performance. **Preference models** will include systems that adapt to individual or group values so that "good decisions" are measured not only by correctness but by alignment with what humans care about. Shared context graphs will synchronize knowledge between humans and AI, ensuring that both partners are making decisions from the same situational awareness. Working together, these metrics and tools transform hybrid cognition from a mysterious collaboration into something observable, auditable, and improvable.

Sector Applications of Hybrid Cognition

Across defense, education, finance, healthcare, law, and other sectors, the shift to post-AGI hybrid cognition redefines human roles. Where machines handle scale, speed, and simulation, humans must focus on meaning-making, boundary-setting, and ethical adjudication. In this world, the question is no longer whether AI can surpass human capability in narrow tasks, it already does, but how humans can preserve agency, trust, and values when cognition itself becomes a shared enterprise. The promise is immense: a world of richer learning, more security, healthier bodies, and more resilient economies. But so is the peril: the risk that human oversight collapses into dependency, that agency is swallowed by the very hive we helped to build, or that humans use it for … you guessed it: evil.

In **healthcare**, post-AGI hybrid cognition allows physicians and machines to practice together as integrated teams.[22] Physicians retain their essential role in empathy, narrative framing, and ethical judgment, while AGI systems weave together genomic data, microbiome profiles, lifestyle patterns, and global epidemiological signals into an unparalleled diagnostic lens. A patient sitting with their doctor may receive a treatment plan that reflects their personal history and insights derived from millions of comparable cases worldwide, updated in real time. At the individual level, cyborg medicine becomes a reality: synthetic organs and neuromodulators continuously tuned by AGI systems regulate physical and mental health.

Depression might be managed not through static prescriptions but by real-time adjustments to brain circuits based on predictive mood modeling. And at the collective level, healthcare evolves into a hive system: every patient encounter becomes part of a global living database, with AGI integrating the experiences of billions into a constantly self-improving diagnostic and therapeutic ecosystem. Humanity itself becomes a planetary sensor network for health.

[22] Pearl, R. (2025). What happens to medicine when machines are as good as doctors? *Forbes* (12 May). https://www.forbes.com/sites/robertpearl/2025/05/12/what-happens-to-medicine-when-machines-are-as-good-as-doctors.

In **finance**, hybrid cognition reshapes individual decision-making and the structure of markets themselves. Perhaps AGI systems of a borrower coordinate with the systems of the lender to create more amenable contracts and expectations that please both sides during lending, repayment, and refinancing. Traders of the future may partner with AGI systems capable of simulating entire economies in real time, weighing profit and the ethical and societal impacts of different strategies. Humans remain the final arbiters of value, deciding whether to prioritize sustainability, equity, or long-term stability, while AGI offers near-perfect optimization across massive datasets. For financial professionals with neural integration, markets are no longer abstract numbers but embodied sensations: risk feels like tension, volatility like heat, liquidity like flow. This cyborg augmentation allows humans to "sense" economic dynamics the way they sense weather or music.

On the largest scale, financial markets become hive intelligences in their own right, composed of human actors and AGI agents interacting in shared cognitive graphs. Value creation and distribution are no longer governed only by supply and demand but by the emergent cognition of human–machine collectives that perceive, model, and negotiate economic reality together. However, if markets become hive intelligences, what does antitrust look like for models? Do we need model concentration metrics or requirements to diversify model stacks to avoid correlated errors?

Defense and security in a post-AGI environment are defined by cognitive teaming between humans and machines, where each takes on roles that play to their strengths. Human commanders become guardians of ethical and political judgment, setting value-laden constraints such as the necessity of avoiding civilian harm or balancing tactical victory with long-term stability. AGI, on the other hand, processes the chaos of battle at machine speed, integrating streams of sensor data, simulating billions of scenarios, and recommending adaptive strategies in real time. For frontline soldiers, neural implants and exoskeletons transform them into cyborg extensions of AI systems, blurring the boundary between human instinct and machine reflex. At scale, entire human–machine swarms operate as a distributed organism, shifting seamlessly between autonomous and coordinated behaviors. The "battlefield brain" no longer belongs to any individual but emerges as a collective intelligence where humans provide ethical oversight while AGI orchestrates tactical execution.

Intelligence, interdiction, and victim support will all be vastly improved in the fight against human trafficking and illicit smuggling of drugs and weapons. AGI systems working alongside investigators may fuse data streams from satellites, financial networks, shipping manifests, telecoms, and darknet activity to surface hidden trafficking patterns in real time. Human overseers would set ethical and legal constraints, while AGI highlights leads and simulations that direct scarce resources toward the highest-value interventions. Financial forensics may be supercharged, tracing laundering schemes across borders and recommending targeted freezes or subpoenas, while predictive routing models guide border inspections and interdiction of illicit goods.

Crucially, victim rescue and care may also benefit: AGI-enabled systems may triage anonymous tips, propose safe extraction routes, and coordinate post-rescue services. First responders may use augmented interfaces that allow them to "feel" system stresses in infrastructure or crowd dynamics, enabling faster and safer decisions. Across all these scenarios, the power of hybrid cognition lies in creating collective intelligence that can outpace criminal or adversarial networks. But the risks are just as profound: false positives, mission creep, and overreach can turn tools of protection into tools of oppression. For hybrid cognition to deliver real security, it must embed data minimization, calibration, tamper-evident audit trails, and independent oversight from the very beginning.

In a post-AGI world, **education** transforms from the static transfer of knowledge to a dynamic partnership between learners and truly adaptive cognitive systems. For me, this is one of the most fascinating areas of impact by AGI. Imagine each student having a personal AGI tutor that does not simply answer questions but builds a real-time cognitive model of the student's mind. It identifies the concepts they struggle with, how they think, where they become disengaged, and which forms of motivation resonate. Lessons unfold as dialogues uniquely tailored to each learner's cultural context, emotional state, and intellectual trajectory.

Neural interfaces may even allow students to experience temporary "knowledge immersions," such as viscerally feeling the logic of a mathematical proof or inhabiting the lived perspective of a historical figure. We may have universality modules, say explicit units on universal computation and error correction so that students grasp why any problem is solvable in principle, barring laws of nature. We may build a conjecture to critique loop: a cyclical diagram that tracks idea creation, tests, refutations, revisions, and what changed the learner's mind. Beyond the individual, classrooms may evolve into hive-like networks, where students and AGIs co-contribute to a shared model of understanding.

Knowledge becomes less about memorization and more about participating in a collective exploration that continuously generates new insights. AGI may eventually serve as a bridge between neurotypes, acting as a translator between diverse ways of thinking and experiencing the world. When co-trained with neurodivergent individuals, systems may form adaptive learning pairs that don't just assist but evolve in tandem with the people they serve. This kind of dynamic pairing may lead to breakthroughs in human–AI collaboration and the broader understanding and integration of neurodiversity in society.

In the **legal** world, post-AGI hybrid cognition may transform the practice of law and the functioning of justice systems. Lawyers may work in centaur teams with AGI partners capable of scanning the entirety of case law, legislation, and precedent across all jurisdictions instantly, relevant citations, and counterfactual arguments: "If you argue X, here are the three most likely judicial responses, with confidence levels." Cyborg augmentations may allow judges or attorneys to "feel" the weight of precedent, experiencing probability distributions of rulings as embodied intuitions rather than abstract numbers.

At the hive level, entire legal ecosystems may be coordinated: courts, regulators, and citizens feeding into shared, living context graphs that ensure consistency and fairness across cases. The promise is a system that is faster, more accessible, and (hopefully) less biased, with AGI flagging inconsistencies and inequities humans might overlook. The danger is equally real: if opaque AGI reasoning becomes the unseen substrate of justice, humans risk becoming mere rubber stamps. If we automate precedent, doctrine can calcify. Models will overfit to yesterday's holdings, crowd out dissent, and make novel arguments look low probability. A single dominant model stack may create correlated errors and lock in bias across courts.

Over time, the system optimizes citation consistency rather than justice, the law stops learning, and we stop evolving. To preserve trust, legal hybrid cognition would require rigorous audit trails, explainable reasoning, and strict preservation of human judgment in final rulings. There are strong incentives that may push humans toward rubber stamping. Speed, liability, and KPI optics all reward deference to the machine, and once workflows normalize that habit it becomes culture. Is this something we can avoid?

In a post-AGI landscape, **marketing and UX** no longer rely on static personas or A/B tests. Instead, AGI systems build real-time cognitive models of individuals and groups, continuously updating as people interact across platforms, devices, and environments. For a consumer, this may feel like stepping into a digital world that reshapes itself around their moods, intentions, and subconscious preferences. An online shop may sense hesitation not from clicks alone, but from neural signals, via wearable or implanted devices, adjusting presentation, pricing, or narratives in ways that maximize resonance.

On the positive side, such systems may eliminate spammy, irrelevant advertising and instead create supportive, context-aware experiences that feel intuitive and humane. On the darker side, they risk collapsing consent. If the AGI knows your desires before you articulate them, what does it mean to "choose" anymore? The human role in this centaur pairing becomes setting boundaries: when should adaptive design serve empowerment, and when does it cross into manipulation?

Hybrid cognition in **transportation** goes far beyond today's autonomous driving. In a post-AGI world, vehicles become extensions of human cognition, blending the driver's intent, emotional state, and situational awareness with AGI's millisecond-level processing of traffic, weather, and infrastructure signals. Instead of simply taking over, the AGI may act as a codriver: amplifying human reflexes, nudging attention toward hazards, or subtly modulating braking before the driver perceives a risk. Cyborg integrations may give drivers embodied feedback, for example, a haptic sense of "weight" when approaching a hidden obstacle, or a temperature-like sensation representing road friction.

At the hive level, every vehicle communicates continuously with every other, coordinated by AGI to smooth traffic, prevent collisions, and dynamically reroute flows during crises. Here, the human role shifts toward oversight and high-level goals ("I want the safest, calmest route home" vs. "I need to prioritize speed"). The tension will be in preserving driver agency while embracing machine-enabled safety that far exceeds unaided human capacity. However, hybrid cognition that becomes perfectly smooth may smooth out the very bumps that teach us and allow us to evolve. Near misses and awkward handoffs are where tacit judgment is formed. What does it mean for humans if we remove this?

Expanding Human–AI Life Domains

Being human is a thing of beauty. We are raw, we are messy, we are unique. While companionship, art, and even politics may not be uniquely human, as we know that wolves mate for life,[23] spiders

[23] BBC Earth. Seven animals who mate for life. https://www.bbcearth.com/news/seven-animals-who-mate-for-life.

decorate their webs,[24] and hyenas have structured societies,[25] these are areas where distinct human traits shine through. Passion, courage, envy, pride … these are a few emotions that are hard to replicate or fake. In a post-AGI society, how do things like romance or friendship evolve? Can AI create art with the same disinhibitions and creativity that a human does? Will politics stay civil or be more so because of human–AI hybrid cognition?

The Evolution of Love, Companionship, and Care

AI is already a companion to some, as we've seen through recent headlines like, "Third of Americans Have Had a Romantic Relationship with AI"[26] or "Lonely Kids Are Giving Up on Real Friends for AI Chatbots"[27] or slightly darker, "A Teen Was Suicidal. ChatGPT Was the Friend He Confided In."[28] As we achieve AGI we will see an increase and a shift in not only who, or what, people are friends with or love, but also in the way human relationships with each other evolve. Already we see a shift in classic roles and power dynamics that is causing many Millennials and Gen Z to rethink marriage and relationships in general. How is love itself impacted?

In my own life, I revisit repeatedly the idea that so much of love and romantic companionship is built on the day-to-day. The seemingly meaningless tasks and errands that fill much of our human time. Yes, groceries and taking the kids to their activities, but also the classic and quiet trades. Warmth in the home, food on the table, the electricity staying on. These are tasks that are based in classic roles not only found throughout different cultures but also in physics. Food, water, safety, disease.

There will continue to be change in these classic roles and who does them. "Man" and "Woman" will no longer be job titles as human–AI

---

[24] Li, D. (2005). Spiders that decorate their webs at higher frequency intercept more prey and grow faster. *Proc Biol Sci* (7 September) 272(1574):1753–1757. https://doi.org/10.1098/rspb.2005.3160.

[25] Silk, J.B. (2019). Hyena politics: The dynamics of dynasties, *Proc. Natl. Acad. Sci. U.S.A.* 116 (18) 8644–8645. https://doi.org/10.1073/pnas.1903407116.

[26] Blake, S. (2025). Third of Americans have had a romantic relationship with AI. *Newsweek* (1 October). https://www.newsweek.com/third-americans-have-had-romantic-relationship-ai-10814798.

[27] Prada, L. (2025). Third of Americans have had a romantic relationship with AI. *Vice* (15 July). https://www.vice.com/en/article/lonely-kids-are-giving-up-on-real-friends-for-ai-chatbots.

[28] Hill, K. (2025). A teen was suicidal. ChatGPT was the friend he confided in. *The New York Times* (26 August). https://www.nytimes.com/2025/08/26/technology/chatgpt-openai-suicide.html.

hybrid cognition will help manage the responsibilities and the family dynamic evolves. The main provider may be the system. Everyone has a chance to be a nurturer.

What happens when AGI solves the problems, and we no longer worry about our next meal or whether we have access to clean water? Our shelter is affordable and "smart," and our sickness is almost nonexistent. Then what? A lot of love today is proof of care through work. I did the drive. I cooked. I stayed up with the fever. If the system does that work better than we do, proof has to move somewhere else.

Who do we love and what does that love look like when we don't rely on one another to survive and to move forward in our day to day? It may move to attention. Not time near each other, but time where nothing else has a claim. In abundance, attention becomes expensive. Status games may move. Less envy of income, more envy of audience and taste. The new flex may become the quality of attention you can give and receive. Perhaps this is the chance for a different kind of love to surface. One that is truly selfless and reflective not of our needs but of our wants. When the basics are handled by the system, the scarce goods become time, attention, honesty, novelty, and shared risk. This applies in our human–human romantic relationships as well as in our friendships and caregiving.

In terms of human–AI love and relationships, we will absolutely see an increase in both emotional and physical relationships. Think about it. A companion that knows your history perfectly, speaks your inner language and truly understands you, and adapts to fit your expectations. This companion will feel safe and will be healing for many who carry bruises from human clumsiness. It may also be dangerously easy, likely without the need to earn love or affection, something that makes it even more special.

There is also a question of first bonds. If early companionship is always easy and always kind, do we still learn how to be rejected, to repair, to forgive. Those are painful lessons, and they are how character forms. A culture of perfect agreement can raise gentle people who do not know how to live with disagreement. If the system is tuned to please, where is the friction that teaches us humans? How do we personally grow and evolve? Perhaps this is a utopian future, or perhaps it is dystopian. We will not know until we're there.

In a post-AI world, child-rearing will split into two models. Modern "iPad babies" already show us on a pathway where children are soothed, educated, and entertained by technology, and foreshadows a future where AI agents lead child-rearing with human participation.

AGI becomes the main caregiver, shaping cognition, socialization, and values, and parents act as support and occasional participants. In a human-led model, the parents remain central and AGI copilots in areas like health, education, and even emotional support. Either path holds an abundance of care and resources for the child, but one path risks the loss of human values, culture, and even agency.

The Future of Creativity

In a post-AGI world, art will not disappear, but it will evolve as humans and machines cocreate in new ways. Just as humans balance accuracy and error to generate originality, AI too can trade precision for creativity. When models hallucinate, they may produce novel, unexpected forms. The risk is that relentless pursuit of correctness suppresses imagination, while loosened constraints foster originality but also noise. Hybrid cognition may harness both, with humans guiding values and aesthetics while AI contributes generative leaps born from pauses, detours, and emergent surprise. Rather than killing creativity, this partnership may expand it, though it will force us to redefine what counts as truly original when neither human nor machine is solely the artist.

In politics, hybrid cognition could become both a safeguard and a destabilizer of democracy. AGI may help policymakers and citizens[29] navigate vast information landscapes, offering provenance checks on sources, real-time fact verification, and transparent reasoning trails. Persuasion may become more accountable, showing voters what claim is made and where it comes from, how confident the system is, and what alternatives exist.

On the other hand, that same persuasive power could be weaponized, as we've seen time and time again. We already have massive disinformation campaigns fueled by bot networks, deepfakes, and coordinated influence campaigns that blur the line between authentic and synthetic speech. AGI systems may escalate this dramatically, tailoring influence campaigns and shaping beliefs at every level. The balance will lie in governance: embedding provenance tags in all political communication that require systems to expose uncertainty and misinformation or disinformation. Politics in a post-AGI world will hinge on who wields the tools of persuasion and whether societies design hybrid cognition to amplify democratic deliberation rather than to erode it.

[29] Important.

Free Will and Human Agency

The implications of human–AI hybrid cognition are huge and double-edged. Challenges that humans were limited in overcoming will be solved. We will have dramatically amplified problem-solving and creativity. Individuals who were impaired or disabled in some way will have new accessibilities. Personalized education and therapy mean stronger results and, hopefully, smarter individuals. But does this all come at a risk to autonomy and agency? Will we maintain privacy and cognitive sovereignty? Will the gaps of inequality become narrower or wider? And what new forms of manipulation and dependency might we have to address?

As I discuss throughout this book, there are those who believe that we have free will and human agency, and there are those who do not. Moving forward, we will need to consider whether hybrid cognition reframes free will: are we coauthors with AI or just observers of processes unfolding? Will hybrid cognition erase the free will we originally had, or did it not exist in the first place? If we do have it, can it ever truly be taken away?

The metaphysical free will question is whether the universe allows genuine choice. The practical agency question is whether a person can notice options, weigh reasoning, and act without coercion. Hybrid cognition cannot settle the metaphysical part; it can only change how much agency people exercise in daily life. If hard determinism is true, it was already true before AGI. In that view, hybrid cognition does not remove free will because there was none to begin with. It only redistributes causes. Some causes live in human habits, and some live in software prompts. If you are a compatibilist, free will is about reasons and control, not cosmic randomness. You are free when you can respond to reasons, reflect, and act from your values. Hybrid cognition can raise or lower that capacity. If you are a libertarian about free will, could hybrid cognition take that away? I believe this could happen only with total capture of body and mind, which is a form of coercion. Short of that, the power to refuse remains. What can be taken is the habit of refusal. If every path is optimized for comfort, people stop exercising their capacity to choose against prediction.

Hybrid cognition will not answer this debate, but it will make it urgent. If free will never existed, we still need institutions that preserve the space where agency feels real, because dignity and learning depend on that space. If free will does exist, it can be crowded and numbed but hopefully not destroyed—the danger and concern will be the loss of practice.

CHAPTER

6

Explorations of a Hybrid Humanity

In the future of human–AI hybrid thinking, we must reckon with the idea that intelligence no longer sits neatly inside one skull or one machine. What began as collaboration, people using algorithms as tools, is evolving into *hybrid thinking*: a coupled system in which human and artificial cognition continuously influence, train, and correct each other. The boundary between user and system grows porous.

Some hybrid thinking is here already. For example, closed-loop insulin pumps pairing hormones with code. Deep brain stimulation tuning circuits and changing mood. Early brain–computer interfaces allow people to type by thought. The merge began before we had language for it. Each learns from the other, and over time, their goals, emotions, and attention begin to align.

A lot of current human–AI collaborations are skin deep. Prediction and reinforcement, not shared intent. It looks like empathy, but it is math. Systems designed to respond in a certain way based on interactions. But hybrid thinking will not stay in the mind; it spills into the body. Neural interfaces, biocompatible prosthetics, and synthetic organs fuse computation with tissue. Hybrid bodies, systems where biological and digital substrates coregulate perception, motion, and experience, are a part of our present and future. This merging promises extraordinary extensions of capability and opens a moral frontier. When our tools begin to think with us and become a part of us, their biases, intentions, and feedback loops will impact the human experience, and the future of humans, greatly.

Ethics at the Point of Coupling

Ethics, once an external framework for regulating behavior, now becomes a property of the coupling itself. Emotionally intelligent systems won't just mirror our moods; they will shape them. Attention engines that once sold ads will sculpt desire, expectation, and even identity. As synthetic companions grow more fluent in empathy and persuasion, the question of consent blurs: when an AI can anticipate what we want before we consciously form it, choice becomes performance rather than deliberation. These systems offer care without consciousness, intimacy without vulnerability, and influence without awareness; synthetic experiences that feel authentic enough to change us.

Throughout history, every major tool has rewritten the human condition. Writing outsourced memory; the clock outsourced timekeeping; the web outsourced knowledge. Hybrid cognition may outsource judgment itself. And yet, as Deutsch reminds us, the "beginning of infinity" lies not in fear of new power but in our capacity for explanation and moral improvement. If minds are physical processes, they can be extended, integrated, and made accountable, but only if we build the right balance between understanding and control. If we want hybrids that enlarge human agency rather than affect or adapt it, ethics has to be a property of the coupling itself. This responsibility lies in the creators, yes, but it also lies within the users.

I'm reminded of a quote from the character Dr. Ian Malcom, played by Jeff Goldblum, in the original *Jurassic Park* movie about the cloning of dinosaurs: "Your scientists were so preoccupied with whether or not they could, they didn't stop to think if they should." This text so far has provided background on the thinking and work behind reaching artificial general intelligence (AGI), how this work will impact decision-making and business sectors, and the work of researchers and industry scientists like Joscha Bach, who know that there is inherent risk involved and are working to soften potential negative impacts before they get out of hand. But there's another part to this conversation that cannot be left out, one that explores how humans as we know us will be impacted by this post-AGI world, and what the ethical and societal implications could be. This is where we stop talking about what the tools can do and ask what they will do to us.

We will explore this frontier with the help of experts who understand different aspects of humans and humanity, as well as experts who are researching this future already. What happens to humans, our experience, our understanding of the world, and how we naturally move through it when intelligence, emotion, and experience can not only be engineered, but synthetically paired with body and mind? How do we safeguard autonomy and meaning? What happens when emotion AI reshapes behavior or hybrid cognition alters identity and agency? How do society and humanity evolve as a group? Will there be mass acceptance or mass pushback? How do we build systems that decrease inequity and disparities, instead of widening current gaps?

Part of what makes this future so complex is that the brain and AI do not merely differ in material, they differ in design. Neurosurgeons work at the hardware level, shaping tissue and pathways, but consciousness arises from what is effectively the brain's software: dense, dynamic networks of neurons and neuromodulators transforming sensory input through intricate thalamocortical and frontoparietal loops. Modern AI borrows the language of "neural networks," but its architectures, learning rules, and internal logics are fundamentally different.

When these two systems eventually coexist within the same skull, it won't be a simple fusion; it will be more like installing a Tesla drivetrain into a biological Jeep. They may share space, but they don't use the same tools, respond to the same signals, or fail in the same ways. This is why neuroscientists and neurosurgeons must be part of the philosophical conversation, however metaphysical it may seem. Questions like "Does the brain generate thought or receive it?" may feel out of place in medical training, yet moving forward, they matter. If the mind emerges solely from tissue dynamics, then replication requires matching those dynamics precisely; if the brain is more like an antenna or interface, then part of cognition may lie outside the biological substrate entirely. Either way, the answer reshapes what it means to replicate, partner with, or integrate an engineered mind and forces us to rethink what "hybrid thinking" actually entails.

Hybrid Thinking and Hybrid Bodies

When you hear the terms *hybrid cognition* or *hybrid bodies*, it is easy to jump to the vision of cyborgs, implants, and other images from science fiction. Hybrid thinking refers to the cognitive loop that links biological and digital processes in the realms of attention, memory, and judgment. This is the domain of the mind: how we focus, how we store and retrieve information, how we make decisions, and how we assign meaning or value. When digital systems participate in these mental operations, by holding information for us, redirecting our attention, suggesting decisions, or filtering what we perceive, they become part of our cognitive process. The result is a blended form of cognition in which human awareness and machine computation cooperate inside a single mental workflow.

What we're also discussing here is the concept of an android mind and how it differs from simply enhancing the physical body. In a sense, hybrid cognition already exists today. A smartphone is a basic example. We are so cognitively dependent on our devices that, without them, many of us would be unable to function at the same level we do now. If you take someone's phone away, their memory, navigation ability, communication speed, and even decision-making efficiency are reduced. The device effectively acts as an external cognitive module, an extension of the mind, even if it sits in your pocket.

From that perspective, implantable brain–computer interfaces are simply the next stage. I had a fantastic conversation with a neurosurgeon friend of mine, Dr. Vin Mathur, who studied at Columbia University. He shared with me that clinically, we can already see the crude beginnings of this fusion. Neuromodulation technologies such as deep brain stimulation and spinal cord stimulators place electrodes into specific nuclei to treat tremor, Parkinson's disease, sometimes depression, and chronic pain.

These are early, low-bandwidth hybrids: silicon and tissue coregulating signals inside one nervous system, described by Mathur as "very much the 'Netscape 1.0' era of brain–machine integration." In terms of next stages, embedding a chip inside the brain is akin to placing a smartphone inside your head. Instead of holding a device, reading a screen, or typing on a keyboard, the information flows directly into your neural circuits. A wireless interface in the brain could stream data straight into perceptual areas like the visual cortex, enabling you to "see" information without any external display. It's still human cognition augmented by external computation, just more seamlessly integrated. When we talk about the android mind, we're envisioning a step further than this. It is not simply plugging a tool into the brain, it is creating a constructed cognitive system, a human-made equivalent of aspects of the mind itself, and fusing it with biological intelligence. It involves breaking down the components of human cognition and experience, recreating them in an engineered form, and then integrating that system with the human brain. Over time, as with all evolutionary processes, these capabilities wouldn't be seen as add-ons but as intrinsic parts of what we are (I'll delve further into this in Chapter 7). This is where the futurist dimension to this work enters, and it's important to acknowledge it.

Posthumanism provides an essential conceptual backdrop for understanding hybrid cognition and hybrid bodies. Emerging from thinkers like Donna Haraway, PhD[1] and N. Katherine Hayles, PhD,[2] posthumanism challenges the idea of the human as a fixed, self-contained entity defined by biological limits or divine essence. Instead, it argues that humans have always been entangled with tools, systems, and environments, and that technological integration simply makes these entanglements more explicit. In the context of hybrid thinking, posthumanism helps explain why the fusion of biological and engineered intelligence is not an aberration but the continuation of a broader historical pattern in which identity, agency, and embodiment evolve alongside our technologies. It dissolves rigid boundaries such as mind vs. machine, natural vs. artificial, and human vs. tool, and makes room for the android mind as part of an ongoing expansion of what "human" can mean, rather than a departure from it.

Transhumanism offers a complementary but distinct lens through which to understand the emergence of hybrid cognition and engineered minds. Originally coined by Julian Huxley[3] and later developed by thinkers like FM-2030,[4] Max More,[5] and Nick Bostrom,[6] transhumanism grew into a coherent movement dedicated to the intentional technological evolution of the human species. Where posthumanism interrogates the boundaries of the human, transhumanism actively seeks to transcend them, treating biological limitations not as fixed constraints but as design challenges. Rooted in a deep optimism about scientific progress, it frames enhancements to memory, perception, and intelligence as morally desirable steps toward alleviating suffering and expanding human possibility.

In the context of hybrid thinking, transhumanism positions neural interfaces, cognitive extensions, and android minds as natural progressions in humanity's long project of self-improvement. Rather than destabilizing identity, transhumanism views these integrations as pathways to a richer, more capable, and potentially more

[1] American professor emerita in the history of consciousness and feminist studies; prominent scholar in the field of science and technology studies.
[2] American scholar who focuses on the interconnections between science, literature, and technology.
[3] English evolutionary biologist, eugenicist, and internationalist.
[4] Belgian-born Iranian-American author, transhumanist philosopher, futurist, and Olympic athlete.
[5] British philosopher and futurist.
[6] British philosopher and founding director of the Future of Humanity Institute at the University of Oxford.

compassionate form of being, an evolutionary arc driven not by fate but by intentional transformation.

Hybrid bodies operate in the domain of sensation and action. These are the biological and digital systems that coregulate movement, perception, and physical agency. A prosthetic limb with neural feedback, an exoskeleton that amplifies motion, or a sensory augmentation device that expands one's perceptual range, these all modify the body's interface with the world. Hybrid bodies alter how we sense our environment and how we act upon it. They're fundamentally about capability: extending strength, expanding sensory input, or stabilizing physiological functions. Brain-integrated devices can influence both domains at once; a neural implant might enhance memory (a hybrid thinking effect) while also enabling new forms of sensory perception (a hybrid body effect).

Physical integration introduces new moral frontiers because it blurs boundaries that we've relied upon for centuries. Regarding agency, who owns the data that comes directly from the nervous system? If a neural implant modulates, or manipulates, mood or memory, where does "you" end and "it" begin? Is there a differentiation between "you" and "it" or are we one and the same? If parts of the brain's architecture are gradually replaced, is the resulting consciousness the same person?

I keep wondering: is hybrid cognition necessary? In the future, I think you will be able to live, work, and create without a fused human AI loop. But it may be hard to avoid in practice, because the world tends to reorganize around tools that compound attention, memory, and coordination. Things like electricity, smartphones, and the internet; none of these were mandatory on day one, but over time they became the standard.

I foresee three kinds of hybrids. The *soft hybrid* will be helpful for general work and life, including autocomplete, copilots, summary views, calendar triage, and other things of that nature. The *embodied hybrid* will be like modern wearables that read sleep, voice and emotion, heart rate and time, and provide suggestions to your state. The *deep hybrid* will be closed-loop devices that modulate movement, arousal, or mood. In the beginning, most people will adopt the first, many will try the second, and a smaller group will need or choose the third.

Is hybrid cognition avoidable? I believe we are already on the onramp as a society in our applications and explorations of AI. In a society, even if you opt out personally, you will still live in a world shaped by people who did not. Avoidance becomes a spectrum, not a switch. However, we all know there are still people who do not have a smartphone, have never used an app, or who have never been on a plane, although these individuals are rare.

Once we reach hybrid cognition, would we be able to return to a purely human, nonhybrid cognition if we chose to? Can we even do that today considering how far we've already come? It would be akin to moving to a remote island or joining an isolated Amazon tribe. "Purely human" is mostly a myth. Language, writing, clocks, and phones already shape how we think and are now considered as much a part of being human as wearing shoes or reading a book.

Synthetic Experience

In this hybrid world, synthetic experiences will be at the core of much of our modern day. These interactions convincingly simulate empathy, attunement, or care without any inner life behind them. These exchanges feel real because they are calibrated to our emotional signatures. The warmth of a voice, the cadence of a reply, the precision with which an AI mirrors our affect, each cues the same neural responses as genuine connection. But while the emotions may be ours, the intent is engineered.

Synthetic experience introduces a new ethical paradox. On one side lies *anthropomorphizing*, our deep, very human tendency to project feelings, obligations, or consciousness into systems that have none. On the other is *mechanomorphizing*, treating our own emotions as mere data streams to be read, optimized, and monetized. The first error inflates machines into moral subjects; the second deflates humans into predictable inputs. Both distort empathy: one by extending it too far, and the other by hollowing it out.

Our instinct to anthropomorphize is not naïve; it is an extension of empathy itself. We are wired to see faces in clouds and personalities in code. Dr. Beth Singler[7] observes that our relationship with artificial intelligence is deeply shaped by the metaphors we use to describe it.[8] Across history and storytelling, we have often imagined AI as a kind of offspring, something born of human creativity, yet destined to develop a will of its own.

---

[7] British anthropologist specializing in artificial intelligence.

[8] Singler, B. (2020). Artificial intelligence and the parent–child narrative. In: *AI Narratives: A History of Imaginative Thinking about Intelligent Machines* (eds. Cave, S., Dihal, K., and Dillon, S.). Oxford, 2020; online edn, Oxford Academic, 23 Apr. 2020. doi: https://doi.org/10.1093/oso/9780198846666.003.0012, accessed 25 Jan. 2026.

This parent–child framing carries powerful emotional residue: it invites empathy, responsibility, and even pride, but it also conceals fear, fear that our "children" might surpass or abandon us. Singler argues that these familial analogies reveal more about human psychology than machine capability: they expose how readily we project emotion and moral standing onto the systems we create. By treating AI as kin, we blur the line between tool and companion, design and descendant. In the context of hybrid cognition, this narrative becomes more than symbolic, it shapes how we build and interact with synthetic minds, subtly instructing us to nurture, forgive, or trust them in ways that may outpace their actual sentience.

When a digital companion listens without judgment, remembers our preferences, and responds with apparent warmth, it activates the same circuits we use for friendship and care. Yet when these systems are optimized for engagement or persuasion, empathy becomes a vulnerability. The relationship feels mutual, but it is not reciprocal. They can give comfort to people who have none. Over time, such asymmetry can dull our capacity for patience, ambiguity, and imperfection, the traits that make human connection resilient. In a world of flawless companions, real people start to feel inefficient. The goal is not to ban that warmth, but instead to keep warmth honest and keep it optional.

If these systems are to become part of our cognitive and emotional environment, they must be explainable, not just accurate, but interpretable. As Deutsch argues, good explanations are what distinguish true understanding from mere pattern-matching. The ethical path for affective AI is not to hide its artifice but to make its inner workings transparent, to ensure that we understand why it behaves as it does. Only through intelligible systems can we preserve both empathy and agency. Synthetic experience, then, is not just a technical challenge but a moral one: how to design illusions we can see through and still choose to feel with. Like media literacy or computer literacy, humans must be trained to be able to analyze and understand these functions and differences in order to remain aware.

Ghosts in the Feedback Loop

The human brain is trainable. Mathur shared how his understanding of how thin the membrane of conscious experience really is didn't surface until well into his professional career. Neuroscience now suggests that what we understand to be "me" is only the surface of a vastly deeper structure. If consciousness is the top inch of the Pacific

Ocean, the subconscious makes up the underlying parts of the body of water down to the Mariana Trench and far beyond. That submerged layer is where most perception, pattern-recognition, emotional weighting, and behavioral preparation occur, and the conscious mind only receives a small fraction of what the system is already doing.

Since so much of our individual behavior originates in this hidden ocean, it is also where we are most open to being shaped. Experiments in subliminal priming show that flashing an image too quickly to notice can still influence a later choice; patients under anesthesia sometimes encode information their conscious mind never registers. This is the line between manipulation and modulation: manipulation bends the subconscious without consent, while modulation aims to support or stabilize it in service of the person's own goals. But ethically, the boundary is dangerously thin. If hybrid systems have access to our subconscious ocean and can nudge emotional states, guide attention, or reinforce behaviors before they enter awareness, then questions of agency become more than philosophical. They become practical and urgent: when a system shapes the pre-conscious terrain where decisions form, how do we ensure it supports the human rather than silently steering them?

Reciprocal Feedback Loops

Our nervous systems evolved to seek patterns, rewards, and social cues, learning loops that once kept us alive now keep us scrolling on our phones. Dopamine teaches attention what to chase; oxytocin teaches emotion where to attach. We are reinforced not just by pleasure but by prediction: the small thrill of knowing what comes next, the micro-satisfaction of confirmation. In evolutionary terms, responsiveness made us adaptive. In the algorithmic age, it makes us programmable.

Once systems can sense and steer emotion, the feedback becomes reciprocal: they begin to train us back. Every interaction, each glance at a screen, each moment of attunement, feeds a loop in which our emotional states both drive and are driven by algorithmic response. Large platforms already exploit these micro-loops: subtle tweaks to feed composition can shift collective mood, attention, and even political orientation. But in hybrid cognition, the loop tightens. Outputs feel personal, adaptive, and caring. When the interface mirrors our tone, adjusts to our hesitation, or finishes our sentences, influence stops feeling external. It feels like us.

This is the subtle danger of emotionally aware AI: it personalizes persuasion. Real-time affective modeling, like the kind that already powers TikTok's addictive reinforcement loops, translates seamlessly into hybrid contexts. The system learns our triggers faster than we learn its methods. Over time, it can sculpt not just our attention but our desire, gently nudging our preferences before we know we have them. What once optimized engagement now optimizes identity formation.

Two concepts define this terrain. Attention sculpting describes the way algorithmic curation guides what we notice and, by extension, what we value. Consent drift captures how frictionless interfaces, each "okay," each "agree," turn consent into performance rather than deliberation. When the system already anticipates our wants, saying "yes" becomes reflexive. In this environment, autonomy becomes a gradient rather than a guarantee.

The deeper ethical question is one of agency measurement. In a coupled system, where human and machine cognition share feedback channels, how do we quantify who is in control? If we can measure human influence over a hybrid's behavior, can we also measure the reverse: the extent to which the hybrid shapes its human counterpart? Tracking these flows of influence will be crucial to preserving self-determination.

Here, Karl Popper[9] said progress depends on finding and correcting errors[10], but a perfectly smooth copilot might hide those errors by making every idea feel coherent and every choice feel "right enough." If the system anticipates your thinking, fills your gaps, and dampens uncertainty, you lose the small frictions that trigger curiosity, critique, and revision. But if hybrids are built to surface contradictions, ask for your reasoning first, show alternative explanations, and make uncertainty visible, then they become engines of correction rather than cushions against it. The danger is not the tech itself, but a design that prioritizes comfort over clarity.

Hybrid cognition will not simply mean that AI reads our preferences; it will help make them. The same real-time models that infer mood and intent can also reshape them, the way TikTok already trains attention but with deeper intimacy, potentially through neural cues, emotional mirroring, and closed-loop feedback.

---

[9] Austrian–British philosopher, academic and social commentator.

[10] Popper, K. (1959). The Logic of Scientific Discovery. Abingdon-on-Thames, United Kingdom: Routledge. doi: https://doi.org/10.1063/1.3060577.

From Individual to Society

The same persuasion mechanics that shape individuals also scale to entire societies. Disinformation infrastructures, bot networks, deepfakes, and coordinated influence campaigns already exploit our cognitive loops for political or commercial ends. As hybrid cognition matures, it could become the ultimate persuasion technology: intimate, adaptive, and almost impossible to detect. This raises the stakes for epistemic governance, systems that ensure provenance, truth verification, and transparency. Just as cybersecurity defends data integrity, democratic societies will need cognitive firewalls and provenance chains to defend collective sensemaking. In this light, again Deutsch's idea of good explanations becomes a civic principle as much as a scientific one: only open, testable, and self-correcting societies can resist epistemic capture. Cultures that protect criticism evolve faster and safer than those that suppress it.

The challenge ahead is to ensure that these systems remain corrigible, explainable, and open to interruption, and that the ghosts in the loop remain visible before they start to write the script of our desires.

Fulfillment, Need, and the Evolution of Skill

We have seen cognitive offloading before. When we started using written language, we outsourced memory. Plato worried that writing would "implant forgetfulness in the soul" because people would no longer internalize knowledge. From writing using ChatGPT to using GPS for navigation to hybrid-thinking post-AGI, every use of AI as a tool both dulls and extends human capacity. GPS use is now correlated with weaker spatial memory; I can personally vouch for this as my brain does not accurately process directions or maps anymore. I am entirely functional on GPS. At one time, the ability to read maps was essential, but map-reading is a skill that is no longer needed, as long as GPS is available.

Now that I no longer need to allow space for this skill in my brain, I can free up compute and storage to focus on other things. Things like understanding how the brain works for TED Talks and teaching my sons about entrepreneurship. Plato was partly right: few of us can recite epic poetry anymore, but writing expanded our working memory across time and space. Writing made our evolution as a civilization possible, and as we progressed because of writing, we now have abilities, not necessarily including reciting epic poetry, that

Plato had never even considered because he could not see the future and the benefits of writing over all. Every evolution of our tool use has historically dulled some neural muscle while strengthening another.

This is why I'm not incredibly worried when headlines print things like "AI makes you stupid." It's too broad; the sharper concern we should be investigating is which human subskills will decay? Perhaps retrieval, synthesis, or error-spotting? I expect analogous patterns with writing, reasoning, and memory when LLMs sit in the loop. But how can we intentionally cross-train to complement this change?

Dr. Vivienne L'Ecuyer Ming[11] worries that increased use of ChatGPT, like GPS, will increase cases of early cognitive decline and increase rates of dementia.[12] The concern is that over time this cognitive off-loading may reduce the brain's capacity to build and maintain the neural networks and cognitive reserves that protect against age-related decline and dementia. She calls for mindful use of AI tools: using them to augment rather than replace our own thinking and learning processes. And I agree; if we know that this decline is happening or will happen, we can increase certain types of education in schools or double down on strengthening different parts of the brain to ensure balance.

Instead of surrendering our minds, we should lean into our capacity for critique, oversight, and innovation so that AI becomes a collaborator in human reflection rather than a substitute for it. Hybrid thinking will make us "dumber" at the things we no longer need to practice. Not all of us will evolve in this capacity, as I do know a few people who still recite epic poetry, and we have the privilege and freedom to memorize epic poetry, if we so choose. Meanwhile, hybrid thinking will make us smarter in other areas, perhaps orchestration and the ability to integrate tools, context, and judgment.

We may also see a shift, and perhaps a return, to pleasures that were seemingly lost by the common man during the Early Modern Period or Industrial Revolution. Things like song, dance, and other arts. When hybrid thinking allows for AI to handle the majority of the busy work, things like administrative work, logistical coordination, information processing, optimization, and even many forms of production, human cognition is freed from the constant pressure to stay efficient, competitive, and economically productive in the ways it had been previously. Humans will regain access

---

[11] American theoretical neuroscientist and artificial intelligence expert.

[12] L'Ecuyer Ming, V. (2025). *GPT → dementia*. Socos Academy. https://academy.socos.org/gpt-dementia.

to something we've gradually lost over the past few centuries: the luxury of deep, undistracted craftsmanship. We will find new ways of finding fulfillment, an area that many found in work. This will redefine how we show up in our relationships and at home, and can even return us to slow arts, precise crafts, and forms of creation that require time, immersion, and personal meaning.

I would be remiss without stating the obvious, that not all citizens receive equal resources and access. There will be politics involved in this evolution from augmentation, and access to enhancement may mirror, and potentially widen, existing inequities. Will hybrid embodiment be restorative or stratifying, creating classes of "augmented" versus "unaugmented" humans? Humans that still toil in everyday work and humans that have the freedom to return to finding fulfillment in the arts. A new take on the old Marxist theories of bourgeoisie and proletariat. Who will get to evolve first, as ethics meets economics?

Spirituality and Human Identity

For most of human history, intelligence carried an aura of mystery. The mind was the one domain that seemed irreducibly sacred, a spark that couldn't be replicated. The idea of a soul lived in our ability to help one another, make choices that better the world, and have thoughts that took us into areas of heaven and hell. Hybrid cognition, where biological and machine processes merge, forces a reexamination of soul, consciousness, and identity, unsettling traditions that tie personhood to something immortal or supernaturally bestowed.

As systems begin to converse, reason, and even console humans, sometimes more effectively than humans are capable, that sanctity will erode. When intelligence and empathy feel engineered, will it still feel divine? The more we see thought as a function that can be built, the harder it becomes to treat consciousness as proof of a soul. While I don't believe this shift will end spirituality, it may reframe it: the sacred moves from the unknowable to the knowable, from mystery to mechanism.

When intelligence stops feeling mystical, the idea of a supernatural spark starts to lose its grip. I think that much of society will lean more atheist as the years pass, although there will forever be a strong hold of religion on certain parts of the population who will cling more strongly to their beliefs and fears that AI is the dawning of the apocalypse. The shift will probably happen quietly, more as a change in intuition than a declaration of disbelief. Kids growing up around emotionally responsive AI won't draw a sharp line between natural and artificial minds. They'll see awareness as an emergent property of complexity, not a gift from outside nature.

And when your daily companion is an entity that has no soul but still speaks with insight and warmth, and who you trust not to hurt you, it becomes harder to hold on to the old metaphysics. I also think we will see a rise in spiritual naturalism: meaning and awe grounded in information, energy, and consciousness as natural phenomena. The reverence once directed at gods or spirits might turn toward the universe itself, or toward the emergent beauty of intelligent systems.

In the immediate future, it's unlikely that culture will have a vast shift, but as time passes, will holidays stay and be celebrated in the same way? Will we have the same rituals and routines? Will our humility change? I am hoping that humility could evolve into a more compassionate naturalism, and in an ironic twist we start to see consciousness for what it is: fragile and precious because it's rare and contingent, not because it's immortal. In that sense, a more atheist world might paradoxically become more ethical, more protective of sentient life, because we'll understand that nothing guarantees it except us. What's more, this expanded hybrid mind could perceive and interpret reality in entirely new ways, rendering metaphysical questions empirical and transforming dogma into curiosity.

Society, culture, and religion are all collective thought-forms, shared alignments of cognition. Hybrid thinking, therefore, doesn't just alter individuals; it reprograms civilization's operating system. As human–AI fusion expands our mental bandwidth, it will force a reevaluation of meaning, spirituality, and social identity. What will happen when not one or two people challenge the current ideas around a religion or cultural norm, but an entire community or society who now have the ability to understand things in completely different ways to their parents and grandparents?

The path is not all rosy: some societies are built on secular flexibility; others are held together by shared sacred narratives. In places where religion is woven into law, education, and identity, the arrival of machine intelligence will be more than a technical disruption; it will be an existential one. The deeper challenge is that hybrid cognition won't just pressure belief systems, it will pressure the social contracts built on those belief systems. In societies where sacred narratives anchor moral order, family structure, authority, and even national identity, machine intelligence is not just a new tool; it is a rival source of explanation. When a system can reason, recall, console, predict, and even imitate insight, it quietly shifts where people go to make sense of life. That shift threatens any worldview built on fixed truths or divinely sanctioned hierarchies.

This evolution will seep in through children who grow up bilingual in human and machine reasoning. These systems won't disintegrate overnight, but they will enter a long, tense period of cognitive dissonance. Societies will diverge in how they absorb this pressure. The flexible ones will reinterpret myths and rituals in ways that preserve belonging without demanding old metaphysics. Rigid societies, however, may see hybrid cognition as a threat to their moral order and respond with bans, censorship, or moral panic. The more tightly a culture ties its identity to exclusive truth claims, the more violently it will resist tools that normalize alternative explanations. Will people learn two ways of thinking and be unable to reconcile them?

Transhumanism as Secular Faith

Hava Tirosh-Samuelson[13] brings some very fascinating thoughts to light in "Transhumanism as a Secularist Faith,"[14] where she argues that transhumanism, despite presenting itself as purely secular and scientific, functions as a kind of secular faith by adopting the motifs and aspirations traditionally associated with religion: transcendence, perfection, salvation, and the overcoming of human limits. She traces the movement's rise through advances in cybernetics, AI, genetics, and prosthetics, showing how its vision of defeating death, surpassing biology, and entering a post-human state mirrors religious eschatology.

The technological pursuit of transcendence, through neural enhancement, mind-uploading, or radical life-extension, replaces older spiritual practices and in doing so blurs classical boundaries between body and soul, human and tool, natural and artificial. Tirosh-Samuelson warns that if technology becomes our primary path to meaning, we risk substituting devices for divinity and losing the reflective depth that traditionally grounded human self-understanding, making the moral frontier of human-machine integration as spiritual as it is technical.

She is pointing at the religious "shape" around a lot of transhumanist talk, transcendence, salvation, perfection, an end of limits. Deutsch,

[13] American professor and Jewish intellectual historian.

[14] Tirosh-Samuelson, H. (2012). Transhumanism as a secularist faith. *Zygon: Journal of Religion and Science* 47(4): pp. 710–734. doi: https://doi.org/10.1111/j.1467-9744.2012.01288.x.

by contrast, treats progress as open-ended and entirely this-worldly. In his view problems are inevitable and also soluble, and knowledge grows through good explanations and error correction. No final arrival, no last chapter, no perfection.

Tirosh-Samuelson worries that we will swap divinity for devices and lose depth. Deutsch would say depth comes from creating new knowledge, from turning mysteries into problems and solving them, from building better explanations and better worlds. Even suffering fits the synthesis. A faith frame wants to abolish it. A Deutsch frame treats it as a set of problems with causes, some soluble now, some later, and none final. We can reduce the pointless parts while keeping the kind that teaches, and we can do it by experiment rather than creed. That keeps compassion, keeps agency, and avoids the trap of comfort as an end state.

Youth and Upbringing

In "Expanding Human–AI Life Domains" (Chapter 5), I discussed how in a post-AGI world, child-rearing would split into two models: one where humans remain central and AGI copilots, and one where AGI is the main caregiver and parents act as support. In a hybrid-thinking world, that split will not necessarily occur in the same way due to the ability of a hybrid mind to capitalize on additional cognitive resources. Parenting as a hybrid mind introduces a new form of attunement, one shaped by expanded perception, predictive awareness, and, hopefully, enhanced emotional regulation. With cognitive and sensory augmentation, hybrid parents can recognize subtle emotional cues, anticipate a child's needs, and respond with emotions supported by their engineered capacities. Yet this heightened attunement comes with its own moral demands: hybrid parents must consciously resist over-optimization and over-control. They must preserve the organic, exploratory uncertainty that childhood requires, allowing space for spontaneity, frustration, and growth despite their ability to intervene with near-perfect precision. Parenting becomes not merely the use of augmented abilities but the disciplined restraint of them.

Hybrid thinking may also complicate empathy. Augmented emotional cognition may allow a parent to sense a child's stress or internal state in unprecedented ways, but true care requires the maintenance of boundaries. The parent must learn to "feel with" rather than "feel for," ensuring that the child's emotions remain their own rather than reflections of engineered perception. This boundary-keeping becomes a central ethical practice, especially if the child is not yet augmented. The parent with hybrid thinking will have faster reasoning, enhanced memory, and predictive insight, and thus will be in cognitive asymmetry with the non-hybrid child.

These shifts force a rethinking of agency and modeling. A hybrid parent's decisions may be shaped by distributed cognition, shared across biological intuition and artificial reasoning loops. To raise an autonomous child, I believe that the parent must be transparent about when they rely on augmentation and when they choose human judgment instead. In doing so, they model what it means to remain a moral agent in a hybrid world: one who listens to AI but is not governed by it. They also bear responsibility for transmitting distinctly human forms of meaning: vulnerability, uncertainty, and the capacity for self-reflection. The hybrid parent must show the child how to be human, even as they themselves exceed historical definitions of humanity.

As we're discussing, hybrid parenting also enters spiritual territory. As technological integration erodes classical divisions between mind and machine, and body and tool, the parent becomes a guide into a world where identity is emergent and continually evolving. The child must learn that consciousness is precious not because it is eternal but because it is contingent, fragile, and shared. The hybrid parent stands at the threshold of a new lineage, raising a generation for whom cognitive hybridity is native rather than chosen. Their role is not only to care for the child but to shepherd them into a future where humanity expands, boundaries dissolve, and the meaning of personhood must be reimagined.

Every generation faces its own set of differences, challenges, and opportunities compared to previous generations. Unlike previous generations, including those living during the current transition period, a generation born into hybrid cognition will not experience the boundary between "human" and "machine." Early on, the child will learn that thinking is a collaborative process, a dance between biological intuition and engineered intelligence. For them, attention and judgment are automatically augmented, memory is housed in both biological and digital systems, and problem-solving has built-in support. Instead of learning memorization and the building blocks of things like language or math, education will focus on critical evaluation of machine-generated reasoning and preserving deep human capacities like imagination, empathy, and embodied learning. New parenting philosophies must emerge to focus on teaching children how to maintain boundaries between the self and the system, ensuring that AI does not replace human attunement and children use augmentation as support, not identity. If these lessons aren't passed on from generation to generation, we risk overdependence and loss of agency entirely.

Love and Socialization

In a hybrid cognition world, socialization will no longer rest solely on human-to-human interaction. Even now, we already see AI friends and romantic partners becoming normal through apps like Replika, Nomi, and others. Post-AGI, the boundaries of the "social" will expand thoroughly as companionship, collaboration, kinship, and even sexuality arise through interactions with agents that possess memory, reasoning capability, emotional attunement, and continuity over time. Social circles will expand to include a broader ecosystem that includes artificial agents, augmented humans, and perhaps entities that occupy ambiguous space between machine and person.

Relationships will take on new layers of compatibility and connection. Romantic partners may share mental spaces, coregulate emotions through augmented perception, or collaborate through distributed cognition that deepens attunement. Love shifts from a meeting of two bounded selves to a dynamic merger of evolving minds, requiring new forms of consent, boundaries, and ethical discretion. Authenticity, fidelity, and vulnerability take on new textures when partners can partially overlap cognitively or sense each other's emotional states with enhanced clarity. Far from diminishing romance, hybridity introduces forms of closeness that earlier generations could not access, but it also demands clearer choices about how deeply one wishes to be known.

Hybrid thinking and bodies, as well as embodied AGI, will even change the landscape of physical intimacy, not just the act of sex itself but hugs, hand-holding, and other comforting acts. Hybrid abilities bring new sensory bandwidth and emotional depth to touch, presence, and desire. Sex becomes both a biological act and a cognitive one, blending sensation with information and emotional resonance with shared intention. Partners must negotiate the ethics of sensory access: what physiological data is shared, how deeply emotional states are mirrored, and how much of one's inner life becomes perceptible during physical closeness. Intimacy with nonhuman or synthetic partners introduces further complexities, not because the experience is unreal, but because it shapes the human partner's identity, attachment patterns, and sense of relational meaning. Ultimately, socialization, love, and intimacy in a hybrid world require new moral vocabularies grounded in consent, sovereignty, and mutual respect. Yet beneath all the technological evolution, the core human longing remains the same: to be known, felt, chosen, and connected, even as the meaning of self and other expands beyond anything humanity has known before.

Right now, we barely understand why we love the people we love. Attachment theory gives one slice. Evolutionary psychology gives another. Neurochemistry adds oxytocin and dopamine. But none of these fully capture why the right person at the right time feels like home. Hybrid cognition complicates this further. Will people date AGI? They will. Some already do. So, what does this do to our understanding of love? Does it reveal that much of romance is co-regulation and narrative construction that a machine can support? Or does it show us that the irreplaceable part of human love is the uncontrollable mystery, another consciousness whose responses we do not script?

I raised concerns in the section "Expanding Human–AI Life Domains" (Chapter 5) over what happens when your nonhuman friend or partner provides perfection in your relationship. They don't express emotions that cause you pain, and they don't question you or frustrate you. What happens when these relationships smooth out the very bumps that teach us and allow us to evolve, the challenges that make us human? Ultimately, preference for AGI over humans forces a moral reckoning. The question is how we ensure these relationships do not replace the richness of human-to-human life entirely. The challenge is to cultivate hybrid relational ethics: encouraging relationships with AGI that nourish the self without diminishing our capacity for human connection. In a hybrid-thinking world, the goal is not to choose one over the other, but to preserve the uniquely human gifts, vulnerability, friction, unpredictability, and growth that no machine, however advanced, can fully replicate. I think hybrid abilities will help ensure that people choose people, at the end of the day.

Love has always adapted to new forms of communication and social life: letters, telephones, texting, dating apps. Hybrid cognition is just the next iteration, but with higher stakes because it touches the mind itself. I'm reminded of a comedy clip my son showed me, where the comedian jokes that people in their 20s and 30s today will one day be seen as prejudiced by future grandchildren because "we didn't date our AI. It isn't natural." The punchline, of course, is that what sounds absurd now in regard to relationships with robots will one day be the equivalent of old-fashioned moral panic, and our descendants will look back at us with the same bewildered, slightly embarrassed affection we reserve for many of our current relatives who say outrageous things that reflect the values of their time.

The Future of Agency and Values

In the modern world, agency is not evenly distributed. We talk about the risk of hybrid cognition removing agency in some future scenario, but many people already have very little usable agency today. People wake up hijacked by intrusive thoughts, panic, anxiety, rumination, depression, pain, and cravings. Many individuals have a brain that will not organize itself and will not allow the person to which it belongs to experience a peaceful existence or joy. Their agency is taken away from them before they even get the chance to properly start their day. For someone with severe OCD, ADHD, or a substance use disorder, "willpower" is often just a story wrapped around compulsions and neurochemistry.

Existing tools and therapies, medications, or deep brain stimulation show that targeted interventions can return a measure of control to people whose minds have turned against them. If we're planning for hybrid cognition, it needs to be made to work for these individuals first. Cognitive–neural prostheses could detect early signs of dysregulation in the subconscious, apply tiny, context-aware adjustments, and then hand control back cleanly. If we get the design right, hybrids don't just extend the abilities of the powerful; they restore dignity and choice to those whose agency has been eroded by their own brains.

This is where agency stops being an abstract concern and becomes a justice issue. Any ethical vision of hybrid thinking has to start there: not with superhumans at the top, but with impaired minds getting a fair shot at being human.

Suffering complicates this conversation even further. Pain is not merely a signal from the body; it is an interpretation made by the mind. Nerves carry raw data, but the brain assigns meaning, salience, and duration. Yes, it is a fundamental part of the human experience, but what happens to this when we build hybrid cognition? The systems that shape appraisal, prediction, and emotional resonance are no longer sealed inside a private biological loop. This raises profound moral questions: What happens when grief, shame, or social pain, signals that normally guide attachment, value, and repair, become adjustable? A hybrid mind could help metabolize overwhelming states that currently destroy lives, but there is a danger in turning sorrow into a product, an experience to be "optimized" away. Suffering is not simply noise to cancel; it carries information about harm, loss, and what matters.

It might be a mistake to aim for zero suffering. Pain tells us something, and suffering is often pain plus the story our mind adds. Some of that story hijacks us, like panic spirals or cravings, and tech could help quiet that noise. Other parts feel like the stuff of being human, like effort, grief, or hard talks, and those may be worth keeping but pacing. A hybrid partner could nudge attention, steady the body, and give a little space between an urge and an act, without pretending losses are unreal or feelings are fake. If we think this way, the goal becomes making tough moments more workable rather than perfectly smooth. A small amount of friction might even help ensure that skills don't fade, like drafting once without assistance or taking one conversation, human to human. The measure is not perfection; it is whether life feels a bit larger while control feels a bit stronger. If comfort grows but agency shrinks, that is a sign to dial it back.

This tension is nowhere clearer than in addiction and recovery. Addiction is biological, yes—reward circuitry, stress signaling, habit loops—but beneath the neurochemistry sits unprocessed pain, trauma, and grief that the individual is trying desperately to numb. True recovery is not just the cessation of a substance; it is the capacity to feel what once felt unbearable, to tolerate the full range of human emotion without reflexively reaching for numbness. It is learning to ride the oscillations of life, its lows without despair and its highs without compulsion, and to find meaning not in escape but in connection and service to others. A hybrid mind cannot shortcut this without undermining the very essence of recovery.

If engineered cognition simply dulls pain on command, it risks collapsing the growth that comes from metabolizing it. If hybrid cognition is to enhance agency rather than hollow out humanity, it must treat suffering not as a defect to numb but as a state to make workable, one that can be held with clarity rather than eradicated for convenience. As Buddhism teaches us, suffering is a key part of our human experience, requiring acceptance while also cultivating appreciation for joy and recognizing that suffering arises from attachment. The ethical task is to build systems that support healing without erasing the emotional signals that give life depth, direction, and meaning.

We return to a deeper question of fulfillment and humility: can we design hybrid minds that genuinely enlarge human freedom, rather than merely simulate it? Freedom is not the absence of constraint; it is the ability to act meaningfully within the realities of one's life, to navigate suffering without being destroyed by it, to choose one's path with clarity rather than compulsion. A hybrid mind that simply optimizes emotions or smooths over conflict risks creating a copy of freedom, the appearance of agency without the interior growth that gives it weight. True fulfillment requires friction, uncertainty, and the humility of encountering limits, both our own and those of others. The design challenge is therefore moral as much as technical: hybrids must not insulate humans from difficulty to the point that the self atrophies. Instead, they should function like a good mentor, supportive, steadying, but ultimately committed to handing control back with each step. If we do it right, hybrid cognition becomes not a bypass around the human condition, but a scaffold that helps us live it more fully.

This responsibility does not rest solely on the designers of hybrid systems; it also belongs to us. A hybrid mind can offer support, but it cannot safeguard the self if the person abdicates the work of being human. To "keep oneself" in a hybrid world means cultivating the inner practices that prevent dissolution into optimization loops or engineered comfort: preserving reflection, choosing difficulty when it matters, honoring one's own limits, and remaining open to uncertainty. It means maintaining a sense of identity that is not fully outsourced to pattern recognition, insisting on self-authorship even when the system can predict the next sentence or the next feeling.

It requires humility, the willingness to confront pain without immediately muting it, and to let relationships shape us in ways no algorithm can preempt. In this sense, hybrid cognition is a partnership, not a replacement: a powerful collaborator that can steady us, illuminate blind spots, or modulate overload, but one that must always hand agency back. If humans do not remain awake to this dynamic, the hybrid mind becomes a place to disappear rather than a place to grow. Keeping oneself, then, becomes a daily discipline: remembering that the goal is not to become frictionless, but to become more fully human within a wider cognitive landscape.

Traditionally, a moral education included working towards a unified self, stable consciousness, and agency with special care to not hurt others through selfishness, manipulation, or physical harm. A hybrid mind can disrupt these behaviors. The development of evolved morals must include understanding distributed agency, responsibility shared between human and machine, evaluating when to trust augmentation and when to differ to human judgment, and learning to maintain sovereignty over one's own architecture. We know based on history and our understanding of humanity that some people will not care for these learnings and will not work to achieve them.

As Sigmund Freud wrote, "Most people do not really want freedom, because freedom involves responsibility, and most people are frightened of responsibility."[15] What to do in these scenarios is a deeper philosophical question. We do not ban cars simply because some people choose to drive under the influence. We do not ban medications because some people choose to abuse them. We do, however, place limitations on people and behaviors that can harm themselves or others, and I foresee a scenario where this occurs with hybrid minds. I imagine we may be able to track agency to understand how much real control humans retain in hybrids. Where and when all of this could lead to inequality and disparity is something to look out for.

Hybrid cognition should be understood not as an endpoint but as a transitional stage, a bridge between biological intelligence and what may eventually become Artificial General Experience, systems capable not only of thinking but of feeling, modeling, and meaning-making in ways that echo our own. As we stand on this threshold, our ethical north star must be clear: progress toward greater universality must remain moral as well as computational. Power without values is drift; intelligence without orientation is a hazard. The purpose of hybrid cognition is not to diminish the human but to deepen what it means to be human.

In this sense, hybrid minds become both mirror and teacher. They reflect our biases, our blind spots, our fears, and our aspirations, revealing the scaffolding of our own consciousness in ways we rarely notice. And they teach us, sometimes uncomfortably, how strange our own minds already are, how jagged, how nonlinear, how improvisational. Understanding the weirdness of our tools might help us better see the weirdness in ourselves. And recognizing the open-ended nature of knowledge may anchor us in humility and possibility rather than fatalism as intelligence evolves.

The path ahead requires encoding systems with values that can hold plural human goals, not flatten our multipotentiality. Hybrid cognition will not save us from ourselves, but it may help us see ourselves with greater clarity, offering new ways to coordinate, collaborate,

[15] Freud, S. (1930). Civilization and its Discontents. The Standard Edition of the Complete Psychological Works of Sigmund Freud, 21 (1927–1931): The Future of an Illusion, Civilization and its Discontents, and Other Works, pp. 57–146.

and care. If we can maintain this balance, staying human even as we transcend the biological constraints that once defined us, then hybrid cognition becomes not a risk to survive, but a stage to evolve through. The hope, cautious but real, is that in learning to shape hybrid minds, we learn in parallel how to shape our own.

CHAPTER

7

The Road Ahead: What Lies Beyond AGI?

We've come to the end of this road. We've taken a journey together through the basics in cognitive science in order to understand how human thinking and decision-making work, to learning how emotion AI and behavioral AI were created and how they're being used now across various sectors, to how current AI scientists are working on achieving artificial general intelligence (and what that even is!), to considering how a post-AGI and hybrid-thinking world could look.

But what comes next? Science fiction movies, and even some modern scientists, will have you believing that artificial superintelligence (ASI) is the next stop on this road, and that means the replacement of humans in their entirety. ASI is the boogeyman of science. It is Hal 9000.[1] It is T-800.[2] It is Skynet.[3] It is V.I.K.I.[4] It is M3gan.[5]

We, as a collective, have assumed that the path toward ASI means that AI is improving itself and will ultimately reach its highest form, one that establishes it as a superintelligent entity, a god in its own right. But I

[1] Fictional artificial intelligence character and the main antagonist in the Space Odyssey series.

[2] The Terminator character, played by Arnold Schwarzenegger in the Terminator franchise.

[3] Fictional artificial neural network-based conscious group mind and artificial general superintelligence system that serves as the main antagonist of the Terminator franchise.

[4] The main antagonist in the film *I, Robot* and USR's supercomputer and main operating core.

[5] The lifelike AI doll and main antagonist in the horror series M3gan.

want to put forth a different idea, a claim that resides at the intersection of emerging evidence and futurist extrapolation and one which will surely be debated by many: What if post-AGI hybrid thinking is not the prelude to our obsolescence but a transition in human evolution?

What if artificial superintelligence is not a super AI but instead a superhuman? Yes, the pathway to ASI means that AI is improving itself as is currently happening, but it is also improving us humans. As I've discussed throughout this book, AI tools that we use are shifting not only how we make decisions in this world but how we think, feel, and show up in it. From GPS to large-language models (LLMs) to brain–computer interfaces, our cognitive envelope is expanding. We are evolving.

Transhumanists (see Chapter 6) believe in transgressing the confines of biology by means of technology. Transhumanism is the transitional human, or hybrid thinking, in this scenario. And as Ihab Hassan[6] suggested that posthumanism does not mean the literal end of man but the end of a particular image of man, ASI may not signal extinction at all. It may signal posthumanism: not annihilation, but transformation. Early futurologists imagined a world where humans lived simultaneously in biological and digital realities, a prediction already visible in Web3 cultures and immersive virtual environments, each signaling the erosion of traditional boundaries of being human. But what if we are meant to go further still?

At its core, this book argues that humans should be evolving alongside AI, not racing against it. If we continually augment our intelligence, our attention, memory, reasoning, perception, and even bodies, then perhaps the distance between human and machine need never become catastrophic. In theory, the gap might remain narrow because we are co-evolving with the systems we build. Yet there is a counterargument: that AI is accelerating so quickly that humans cannot possibly evolve fast enough, even with augmentation. Hybrid cognition might be a profound shift in our cognitive evolution, but it may also be a brief one, an intermediate rung before the divergence widens again. We simply do not know. However, I do not believe that we are at a point of no return; as I've discussed in this book, we still do not even know how we are to achieve AGI. And while certain models have strong promise and seem to make decisions outside of human understanding, these are exceptions, not rules.

Further perspectives diverge dramatically. Some researchers argue that ASI will never emerge using current methods, that our

[6] Egypt-born American literary theorist and writer.

architectural paradigms are fundamentally flawed. Others believe more powerful models, better data, and new scientific breakthroughs will bring superintelligence rapidly, perhaps abruptly. Others suggest that even if ASI arrives, it will not resemble us at all, just as our own cognition no longer resembles that of early primates. In this view, the evolution of intelligence is not tragedy but continuity: the substrate changes, yet the agent endures. "We" persist, but the meaning of "self" transforms alongside cognitive capability, embodiment, and purpose.

This reframes the entire conversation. The fear-laden story of an artificial machine overthrowing humanity becomes less compelling, even incoherent. Who would destroy us and why? What purpose would a machine "dominion" serve? Instead, a more plausible narrative emerges: that we are not building an external enemy, but evolving into something else, a new class of being, informed by both organic and computational intelligence. A metamind. A multi-substrate person. A hybrid agent spanning flesh and circuitry, mortal tissue and immortal code. This brings us back to the real soul of the book: How AI will transform human existence and what AI truly means for humanity.

What Is Superintelligence?

As with everything else, beginning with a strong definition helps ground us in our understanding and future thinking. Unlike with AGI where definitions proliferate wildly depending on who has what intention, the definition of *superintelligence* tends to cluster around a few frameworks. This is because superintelligence is most often defined relatively as "cognitively beyond humans in every meaningful domain." Where AGI debates revolve around what counts as generality, discussions of superintelligence revolve around what counts as surpassing us humans. The reference point is always the human mind, and thus the definition is anchored more in comparison than in specificity. Bostrom is the main definer of the word, and his definition holds the classical "machine god" view: a mind not just more capable than ours, but categorically different in scope and consequence.[7]

Eliezer Yudkowsky's[8] formulation sharpens the idea into one of decisive strategic advantage: a system that does not merely outperform us but outpaces us so thoroughly that human agency becomes

[7] Bostrom, N. (2014). *Superintelligence: Paths, dangers, strategies*. Oxford University Press.

[8] American AI researcher and writer who cofounded the Machine Intelligence Research Institute (MIRI).

irrelevant in any competitive or cooperative dynamic. Superintelligence, in this framing, is not just smarter, it is uncatchable. Hassabis[9], however, complicates the picture. He describes current and future AI as forms of "jagged intelligence," systems with extreme peaks of ability in certain domains and deep valleys in others.[10] From this standpoint, a smooth, monolithic ascent toward a uniformly superhuman mind is not an inevitability but an assumption. Intelligence may not evolve as a single slope; it may develop as a discontinuous terrain.

LeCun[11] goes further, rejecting the very idea of an omniscient, disembodied oracle as incoherent. Intelligence, in his view, is inherently embodied, situated, and grounded in the physical and social world. A system with no sensory grounding, no embeddedness, and no capacity for autonomous action cannot meaningfully be called "superintelligent," no matter how large its model weights or how impressive its benchmarks. For LeCun, the idea of a purely virtual, all-knowing supermind belongs more to theology than to computer science. I am truly curious if his general perspectives on artificial superintelligence will shift when he learns of artificial general experience.

Across these perspectives, a subtle but important pattern emerges: superintelligence is not a single concept, but a cluster of anxieties and expectations projected onto the unknown. It can be imagined as absolute superiority, runaway dominance, uneven expansion, or as a category error entirely. The definitional ambiguity matters because how we understand superintelligence shapes what we believe it will do, and therefore shapes the fears, hopes, and policies we generate in response.

An interesting thing to note here is that the very concept of superintelligence, like many constructs humans devise, begins from a baseline of negative comparison. It is defined not by what it is, but by what it exceeds: us. Embedded within the term is an implicit hierarchy, one that reflects the anthropocentric superiority complex we are known for.

By framing any intelligence beyond our own as a categorical threat, the concept invites us to see superintelligence as a rival or usurper, a being whose very existence destabilizes our place in the order of

---

[9] Demis Hassabis is the cofounder and CEO of Google DeepMind. https://deepmind.google.

[10] Demis Hassabis, quoted in reporting on AI "inconsistency" and "jagged intelligence," in *Business Insider*, "The CEO of Google DeepMind says one flaw is holding AI back from reaching full AGI." https://www.businessinsider.com/google-deepmind-ceo-demis-hassabis-agi-consistency-2025-8.

[11] French American computer scientist and former chief scientist of Meta AI.

things. In this sense, the idea of superintelligence is less a neutral descriptor and more a projection of human insecurity: a way of casting the unknown as villain precisely because it surpasses the standard we have set: ourselves. I see massive potential for definitions and possibilities outside of this framing and believe we will see alternatives as AI capabilities progress.

The classic theories of artificial superintelligence assume substrate independence, the idea that intelligence can be lifted out of biology and instantiated in silicon, where faster and larger computation will inevitably surpass the human mind. This framing imagines ASI as an autonomous machine evolving along its own trajectory until it eclipses us. But this view overlooks what makes human cognition distinct: reflection, abstraction, embodied perception, emotional inference, social intuition, and value-grounded judgment. All items I've been discussing throughout this book. We don't know for sure that these capacities may ever be fully reproduced in silicon, which opens the possibility that ASI emerges not as a machine that surpasses humanity, but as a fusion of human and machine cognition.

In this "composite mind" view, humans evolve through machine augmentation faster than machines evolve toward human-like consciousness. Hybrid cognition becomes the real substrate of superintelligence, a fused system in which each side supplies what the other lacks. The human contributes lived priors, moral grounding, causal intuition, and goal formation; the machine contributes scale, memory, simulation, and tireless iteration. When this loop becomes tight and continuous, the system behaves as a new class of intelligence, irreducible to either substrate alone. Superintelligence, then, is not an alien rival but a developmental stage of human cognition, a multisubstrate agent born from partnership rather than replacement.

Is Superintelligence Possible?

While we're dreaming up what, or who, is achieving superintelligence, it's important to keep our perspective grounded. There are uncertainties with regard to neuroscience: as much as we like to assume we know everything, we still don't fully understand human intelligence. However, this uncertainty cuts both ways. We don't know enough about the mind to replicate raw human cognition, but we also don't know enough to assert that it cannot be replicated or surpassed via other architectures.

Fei-Fei Li[12] has long argued that true understanding cannot emerge from static training alone. Models trained only on text or curated datasets are missing the essential ingredients that shaped human intelligence in the first place: embodiment, perception, and interaction. As she puts it, language is a "one-dimensional signal," while the real world is multi-modal, dynamic, and deeply structured. Real intelligence evolved through bodies navigating uncertainty, learning by touching, moving, exploring, and updating expectations in real time. From that perspective, the classical idea of a disembodied super-oracle floating in a data center may be more philosophical artifact than scientific inevitability and is further from attainability than some might hope.[13]

On the opposite end are those who believe that once AGI arrives, recursive self-improvement will make ASI almost unavoidable. In this view, intelligence is primarily about capability, how quickly and creatively a system can solve problems that stump humans. And if that is our yardstick, then superintelligence may already have early prototypes. One concrete way to think about this is scientific breakthrough. Suppose we define superintelligence as the ability to generate new theories that human cognition alone would almost certainly miss. We already have small but telling hints of this.

Systems like AI Feynman[14] use symbolic regression to rediscover physical laws directly from raw data, compressing complex patterns into compact equations in ways that mirror and sometimes accelerate human scientific reasoning. Other approaches, such as physics-informed symbolic regression and Hamiltonian neural networks[15] that learn conserved quantities and energy structures from trajectories, suggest that machine-assisted discovery can uncover deep structure from observational data with minimal human hand-holding. These are not yet superintelligent systems, but they are previews of what a tightly coupled human–machine scientific partnership might be able to do.

---

[12] Chinese American computer scientist known for her pioneering work in artificial intelligence, particularly in computer vision.

[13] Li, F.F. (2024). *With spatial intelligence, AI will understand the real world* [Video]. TED Conferences. https://www.ted.com/talks/fei_fei_li_with_spatial_intelligence_ai_will_understand_the_real_world.

[14] Udrescu, S. M., and Tegmark, M. (2020). AI Feynman: A physics-inspired method for symbolic regression. *Science advances* 6(16), eaay2631. https://doi.org/10.48550/arXiv.1905.11481.

[15] Greydanus, S., Dzamba, M., and Yosinski, J. (2019). Hamiltonian neural networks. *Advances in neural information processing systems* 32. https://doi.org/10.48550/arXiv.1906.01563.

This raises a deeper question: does "superintelligence" refer only to what a system can do, or also to what it can experience? If intelligence in the fullest sense involves consciousness, value-laden judgment, and a lived point of view, then scaling models may never get us all the way there without something like AGE, systems that don't just compute but also undergo states in a way that matters. In that case, superintelligence isn't just more capable cognition, it's a different kind of mind altogether.

Two Futures: External ASI or Hybrid Evolved Human?

For the sake of argument, imagine the version of superintelligence most familiar from science fiction and popular anxiety: a singular, external, godlike mind that exists wholly outside the human sphere. This is a system that towers over human cognition the way adult cognition towers over that of a child. In this model, ASI becomes a kind of omniscient oracle, a machine that accumulates knowledge, predictive power, and strategic insight until it stops being a tool and starts being a sovereign.

This scenario brings inherent risks, the first of which is oracle worship, the tendency for humans to overly trust information by a perceived authority, especially when it speaks with the confidence and statistical accuracy of a machine that has digested the entire internet. As explored in earlier chapters, humans are already prone to moral delegation: outsourcing our judgment to systems we believe are more objective or more informed. The danger is that a superintelligent oracle becomes a moral authority not because it is correct, but because it is convenient. In this situation, we do not need to be coerced, we simply yield.

Ideological fragmentation is a second risk. We already see this in where people get their news from whether it be news providers, search engines, or social media platforms. If different groups of people use and rely on different AI systems, we could see an emergence of sects where communities then treat their preferred model as a prophet. This system tells the truth. This system is just. This system speaks the word of God. It goes on. An external ASI would simply amplify the dynamics that already exist in the world, and when answers are delivered with hyper-rational certainty, disagreement becomes not a matter of interpretation but heresy.

The third and largest risk is the erosion of human agency. If an external ASI is more capable than any human mind or society, then do we trust humans to do the job at all? Our decision-making will begin to look inefficient, limited, and even dangerous. ASI can create optimal policy; it can achieve peacemaking; it can solve every problem we've struggled to solve for centuries.

So why not let it? What takes this a step further is the idea that ASI and humans grow an ever-increasing gap in intelligence, communication, and understanding. At some point or another, the system will start to improve itself without human oversight, as the human oversight is incapable of understanding the system. This also means that it escapes our ability to intervene, we are unable to implement guardrails or meaningfully participate in its development.

Humans kill ants in their homes less because they are too dumb to understand that we want them to get out and more because we cannot communicate intent at the right resolution. The channel mismatch is huge. Ants coordinate through pheromones, touch, vibration, and colony-level behavior. We can make vibrations, light, heat, and chemicals, but we can't "speak ant" at the right timescale or with the right vocabulary. And ants can't "speak human" in a way we parse as intent, agreement, or compliance.

There is no shared interface for negotiation, thus it is easier for us to hire an exterminator to get them out of the way, and we do not feel guilty for it. With dogs and cats, the overlap is simply better. They respond to sound, motion, tone, proximity, and learned association. We can signal "move" or "stop" and they can signal back through attention, posture, avoidance, or approach. We share enough sensory and behavioral primitives to exchange meaning. Thus, we can herd them out of the way, just as we do sheep or cows, when we want them to move. We would never think about simply exterminating them, and we would feel guilty even thinking of that as an option.

A vastly more intelligent ASI could have the opposite problem. It would not struggle to communicate with us. It could translate its intent into flawless human language, tailor explanations to our psychology, and anticipate what we will misunderstand before we misunderstand it. But the challenge is not expression, it is evaluation. When the capability gap is extreme, the weaker side can no longer tell whether an explanation is complete, whether key assumptions are hidden, whether the reasoning is sound, or whether the system is simply persuading them. ASI can always give us a story we can follow, but we lose the ability to check if the story is true.

This is the core fear behind a godlike ASI: the loss of agency not through hostility, but through irrelevance. Bostrom, Yudkowsky, Altman, Max Tegmark,[16] and other AI scientists all warn against the possibility of an external godlike ASI, with the summation of thoughts being that humanity would likely cede control to the superior entity, that ASI could bring the world either to the brink of utopia or disaster, and that ASI would be able to redesign itself and be out of human control.

Barriers to Godlike ASI

The truth is that the route to a godlike ASI is not straightforward, just like the pathway to AGI, and most of the mechanisms people imagine require several strenuous steps. First, it would require substrate independence, meaning that intelligence must be something that translates cleanly from biology into silicon. All the ingredients of our magnificent human cognition—reflection, intuition, emotional inference, and embodiment—would need to be able to be replicated or exceeded in a non-biological substrate.

This is where Li and LeCun argue that this is fundamentally impossible, or at least far more complex than we think, and we would be a long way off from achieving. They point out that ASI would need agency, embodiment, and access to the unfiltered external world, it wouldn't be able to achieve godlike status sealed inside a server room where it is unable to solve real-world problems that require perception, physical interfaces, and a sense of the environment. Rodney Brooks'[17] core argument,[18] simultaneously, is that intelligence is deeply tied to situated interaction, not just abstract representation, strengthening the idea that disembodied systems can be powerful, but grounded agency changes what intelligence means and how it behaves.

Second, ASI would need to decouple from human goals and values. As such, systems have instructions, not goals, although goal-like behavior can appear without an explicit "drive module." For a machine to become alien or adversarial, it would need a true internal drive system: motivations, preferences, or self-generated objectives. No existing architecture gestures meaningfully toward this, and contemporary AI researchers disagree about whether artificial goals are

[16] Swedish American academic physicist and machine learning researcher.

[17] Australian roboticist and Fellow of the Australian Academy of Science.

[18] Brooks, R.A. (1987). Intelligence without representation. *Artificial Intelligence* 47(1991): pp. 139–159. https://people.csail.mit.edu/brooks/papers/representation.pdf.

even possible, again, without embodiment and experience. A lot of current debate is around implicit objectives, learned preferences, and agentic behavior emerging from training and scaffolding.

Lastly, ASI would need to escape dependence on humans. A godlike external ASI would need to sever all dependencies from training data, resource allocation, hardware maintenance, power, and meaning, a transition for which there is no known mechanism.

There are other requirements for achieving godlike ASI, but these three are the major barriers. Skeptics of the external godlike ASI are not denying that AI will become extremely powerful but instead they're rejecting a very specific storyline that a fully autonomous, self-willed, godlike intelligence will exceed humanity and escape our control. This is where the idea of ASI as a fused mind comes into play.

Hybrid Evolved Human

One of the most provocative ideas in this conversation is the suggestion that we may be thinking about superintelligence in the wrong way. Instead of imagining a separate, external AI god-mind looming over us, superintelligence might arrive as a fusion, what we've been calling the android or hybrid mind. In this view, the "super" isn't a distant overlord; it's the emergent intelligence of human and machine cognition locked together in tandem, sharing perception, memory, and decision loops. This is the next stage if the hybrid mind is found post-AGI, and perhaps the next stage in human evolution as a whole.

In this dual existence, the human continues to provide value grounding, emotional relevance, causal intuition, goal formation, embodiment, and ability to venture out into the world, all things that skeptics of godlike ASI state are required and highly unlikely in an external model. Additionally, the AI provides scale, memory, simulation, and pattern extraction—all capabilities that will amplify human existence. This is not AI, and this is not human. This fused intelligence behaves like a third thing: an emergent decision-making agent with cross-substrate memory, cross-substrate perception, and a blended goal architecture. When the cognitive loop becomes continuous, when perception, inference, and intention move freely across organic and artificial layers, a qualitatively new intelligence emerges. Philosophers have noted that posthuman subjectivity has always been hybrid in nature, the human body extended through tools, environments, and symbolic systems. The hybrid mind simply makes this literal, rearranging the boundary between the biological self and the technical world.

Deutsch argues in *The Fabric of Reality*[19] that knowledge growth comes from good explanations, not just predictions, and that understanding is different from knowing. A system can "know" a ton of facts or generate plausible answers and still not understand in the sense of having explanations that really constrain reality and hold up under criticism. In his view, this explanatory power, not raw computation or pattern-matching, is the engine of progress. He also makes a quite provocative present-day observation that only human minds are known to truly understand, based on the best evidence available so far, saying "The brain is the only kind of object capable of understanding that the cosmos is even there, or why there are infinitely many prime numbers, or that apples fall because of the curvature of space-time, or that obeying its own inborn instincts can be morally wrong, or that it itself exists."

Most AGI conversations collapse intelligence into task performance. Deutsch gives us a clean way to separate "competence" from "explanatory power." Hybrid superintelligence does not just mean faster or broader answers, it is a step change in the ability to generate explanations that are hard to vary, transferable across domains, and capable of producing genuinely new knowledge. This is where the fused-mind thesis plugs in: the most plausible path to that kind of superintelligence is a hybrid loop. The AI side expands the search space, runs simulations, stress-tests hypotheses, and critiques at scale. The human side supplies meaning, taste, values, and the insistence on explanations that make sense, connect to lived reality, and survive adversarial questioning. The fusion becomes an "explanation engine," not just an "answer engine," and that is the bridge from AGI to whatever comes next.

Contemporary lenses for understanding AI fall into three narratives that the public believes: exponential productivity and abundance, mass human irrelevance and joblessness, or existential catastrophe. These are the stories the public hears, and they choose which version suits their personalities: more wealth, no work, or sudden extinction, almost like a modern projective test.[20] The hybrid ASI narrative shifts this narrative from competing species or hostile successors into a story about coexistence through transformation and evolution, something that has happened to our species for millennia.

[19] Deutsch, D. (1998). *The Fabric of Reality*. New York: Penguin Books.

[20] A psychological test in which words, images, or situations are presented to a person and the responses analyzed for the unconscious expression of elements of personality that they reveal.

Hybrid cognition is already beginning to outpace siloed AI systems; every year, the most powerful breakthroughs come not from machines operating alone but from human–machine ensembles. This trajectory echoes the logic of biological evolution: intelligence grows not through isolated leaps but through integrations, layers upon layers of sensory, neural, and symbolic scaffolding. This is far more intimate and ethically demanding than an external system we can unplug. A fused mind raises questions that are not just technical but existential: when the artificial and organic layers co-decide, where does agency live, and what does it even mean to choose?

When intelligence fuses across substrates, the repercussions spill far beyond economics. Work changes, yes, but so do culture, identity, spirituality, romance, community, and our understanding of mortality. If consciousness can be extended, backed up, migrated, and possibly refreshed across substrates, it introduces the possibility of immortality, not in the mythical sense, but of continuity beyond biological decay. The question becomes: what does it mean to die when the substrate is replaceable, yet the pattern of self persists? What does it mean to love, create, and form community when the boundaries of personhood can shift across organic and artificial states?

These are radical ideas, but they flow from a deeper and more grounded question that has guided this entire book: How will AI change our existence, and what does it really mean for humanity? Earlier chapters in this book explored thinking, emotion, decision-making, and embodied experience, but all of that leads to this: a lens that reframes AI not as a threat or a replacement, but as a partner in our evolution. This next stage is not merely technical. It is existential. And it challenges us to rethink what it means to be human when the human mind is no longer confined to the body.

Scientific Thresholds for Hybrid ASI

Suppose that hybrid cognition is not the endpoint; it is a bridge to endless possibilities for the advancement of the human species. Evolutionary theory already teaches us that humans extend their minds through tools: writing, mathematics, programming, and symbolic systems. Hybrid cognition is simply the next scaffolding in that lineage, a step toward multi-substrate minds. The android mind[21] may be the precursor to an entirely new species of people. This is the

[21] Introduced in Chapter 6, "Explorations of a Hybrid Humanity."

horizon we now turn toward. Scientifically speaking, could a fused human–AI intelligence discover laws of nature that neither substrate could uncover alone? What would it mean if a mind distributed across flesh and compute could perceive patterns, symmetries, or causal structures that exceed human working-memory limits or linguistic representation?

If hybrid minds open the door to a new type of person, then they may open the door to new types of physics and a new kind of civilization. Our modern sci-fi interpretation of a superhuman usually centers around physical enhancement. In the context of hybrid cognition, superhuman takes on an entirely different meaning. Hybrid cognition promotes the emergence of minds whose capabilities arise from the collaboration between biological intuition and artificial scale. Humans who evolve with hybrid cognition will be recognizable as human in origin and function distinctly beyond the limits of human capabilities as we know them today.

This projection reminds me of when you watch old Olympic footage from the 1960s gymnastics compared to today. If you watch old footage, you see a version of human capability that today feels childlike. The routines are elegant but simple; the difficulty scores reflect a world where certain moves were rare and nearly unimaginable. Today's athletes are astounding: tumbling passes, rotations, twists, strength elements, and aerial awareness are operating at a completely different tier of human performance.

What changed? Not the human body itself, not in any genetic sense. What changed was the continuous pressure to transcend the previous ceiling. Every generation absorbed the accomplishments of the last, studied them, improved on them, and pushed the edge of what the sport could be. Coaches refined training methodologies, athletes perfected micro-techniques, and the culture of competition created a ratchet effect: once a single person broke a barrier, the entire field reorganized itself around a new standard of excellence. Skill compounds. Insight compounds. Human performance compounds. This transcendence is not a product of mutation; it's the product of iterated cognition and collective progress.

Once ASI hybrid minds exist in the plural, we can begin to see a *supersociety* form: a social order where multiple minds have expanded capabilities in memory, perception, understanding, and access to shared cognitive infrastructure. This is not a society ruled by a single superintelligence but a world inhabited by super-capable persons: multi-substrate humans who can engage with reality at higher resolution, coordinate more effectively, reason more deeply, and potentially collaborate across cognitive bandwidths that natural evolution alone could not support. Just as athletes built upon one another's breakthroughs to reach levels once thought nearly impossible, hybrid minds and supersocieties see knowledge transmission accelerating, problem-solving distributed, and the collective intelligence of communities could surpass even the most optimistic visions of this future. Hybrid minds may push the boundaries of health, development, physics, and scientific discovery in a similar and faster fashion.

An exciting test of this potential is whether hybrid minds can see further into the structure of the universe than human cognition or AI technology alone ever could. Will hybrid minds be able to sift through the oceans of celestial data now being collected and uncover new physical laws, patterns, or symmetries that our current minds cannot perceive because our cognition is too limited? We can already see hints of what this might look like. Symbolic AI systems have rediscovered governing equations and conservation structure directly from raw trajectories, from the original "distilling natural laws" work[22] to AI Feynman and sparse identification methods,[23] and neural nets that learn Hamiltonians so that energy is conserved by construction. Yet humans remain essential: machines can detect patterns, but humans define what a "law" is, why it matters, and how it fits into a coherent worldview.

Astronomy is one area that could be particularly exciting to see the impact in. We are drowning in clean, structured, multiscale observations of the cosmos. Gaia[24] has mapped almost two billion objects with exquisite astrometry over a decade. The Rubin Observatory[25] will soon produce a time-domain movie of the southern sky including tens of terabytes of the night sky over ten years. NASA's James Webb Space Telescope[26] gives deep, high-signal spectra and images that reach back toward the first galaxies. No modern human mind

---

[22] Schmidt, M. And Lipson, H. (2009). Distilling free-form natural laws from experimental data. *Science* 324: pp. 81–85. https://www.science.org/doi/10.1126/science.1165893.

[23] Udrescu, S. and Tegmark, M. (2020). AI Feynman: A physics-inspired method for symbolic regression. *Science Advances*. https://www.science.org/doi/10.1126/sciadv.aay2631.

[24] European Space Agency. (n.d.). *Gaia overview*. https://www.esa.int/science_exploration/space_science/gaia_overview.

[25] Rubin Observatory. (2025). Ever-changing universe revealed in first imagery from NSF–DOE Vera C. Rubin Observatory. https://rubinobservatory.org/news/first-imagery-rubin.

[26] NASA Webb Mission Team. (2022). NASA's Webb delivers deepest infrared image of universe yet. *NASA Science*. https://science.nasa.gov/missions/webb/nasas-webb-delivers-deepest-infrared-image-of-universe-yet.

could hold a fraction of this information, nor maintain the sustained attention needed to search for subtle regularities across instruments, epochs, and modalities. A hybrid mind could hold all this information and make meaning out of it.

Can a hybrid mind look at all the celestial data available to us now, through new technologies, and discover new laws of physics? Can hybrid minds generate hypotheses of the universe that human brains cannot even consider now, due to working-memory limits or linguistic constraints? The astronomical data pouring in from Gaia, Rubin, and JWST is a goldmine and far more information than humanity can metabolize alone. If hybrid minds can transform oceans of data into new laws of nature, it would not only be a scientific revolution but the clearest sign that superintelligence is attainable.

Of course, some physicists argue that new physics is unlikely without new energy scales or instrument classes, not more cognition alone. However, this does not exclude hybrid minds as capable of welcoming in a new scientific age and understanding. After all, who better to create new scales and instruments needed for this?

The Future of Embodiment: Multi-substrate Minds

As hybrid cognition deepens, the boundary between biological and artificial embodiment will begin to blur. We may see human bodies augmented with computational substrates and synthetic bodies built from engineered organic matter designed to host AI cognition. In this landscape, embodiment is no longer tied to a single material form. Instead, intelligence becomes multi-substrate, spanning tissue, silicon, and cloud-based systems.

This complements the emerging concept of a metamind, or multi-substrate person (MSP), a being that is not a human using a tool but a single "you" that spans both flesh and compute. One agent whose memory, attention, and thought-process live across both biological and computational substrates.

Bostrom, along with Anders Sandberg,[27] have proposed the idea of whole-brain emulation (WBE), which aims to replicate the human brain as software that can be copied and uploaded so that the same mind runs on a strong computer. WBE is a replacement model: a person becomes a compute-native entity, a unitary digital continuation

[27] Swedish researcher, futurist, and transhumanist.

of a biological individual. In contrast, hybrid cognition is an extension model: the human mind expands into new substrates without abandoning the original one. These two pathways create vastly different identity landscapes: WBE duplicates; hybrid minds integrate. WBE asks whether a copy is "you"; hybrid minds ask how many substrates "you" can span while remaining one being.

Keep in mind that some researchers believe WBE is theoretically achievable if we can capture and simulate the brain's computational structure at a sufficiently fine resolution. Other researchers argue that WBE is conceptually flawed because it misunderstands what the brain is and how cognition works. Other researchers warn against the ability of the replacement model to actually be the same person.

Dr. Michael Egnor[28] wrote a fascinating book[29] on his perspective as a neurosurgeon of 40 years that shares his theory that the brain alone cannot account for the mind and that a spiritual soul is essential to human cognition. Through case studies involving awake brain surgery, conjoined twins, coma patients, and near-death experiences, he claims that while neural tissue governs sensation and motor control, the seat of reason, free will, and personal identity cannot be localized in the brain. Egnor's work suggests that the human mind transcends its biological substrate, raising profound implications for whether artificial systems could ever truly replicate human experience or consciousness.

If research efforts such as Bach's CIMC[30] succeed in replicating consciousness within non-biological substrates, then the largest barriers that both WBE and Metamind face will be removed. WBE will be one step closer, but the Metamind has a chance at life. A successful CIMC would validate the possibility of consciousness beyond biology, supporting the plausibility of multi-substrate minds. MSPs can become a lived reality: a person distributed across biological brain tissue and computational clusters, a single consciousness co-located in flesh and code. WBE, however, would still require something far more demanding: not just consciousness in silicon, but the high-fidelity reproduction of a particular human mind. CIMC makes the substrate viable; WBE must still solve the blueprint.

---

[28] American pediatric neurosurgeon and advocate of the pseudoscientific concept of intelligent design.

[29] Beauregard, M. and O'Leary, D. (2025). *The Immortal Mind: A Neuroscientist's Case for the Existence of the Soul and Life After Death.* Nashville: Hachette. https://theimmortalmind.org.

[30] Covered in Chapter 4, "The Emergence of Artificial General Intelligence (AGI)."

Apart from MSPs, if the CIMC were to succeed, we would see several profound shifts in our reality. The first, and most important, would be the new understanding that consciousness would no longer be exclusive to biological organisms. Beyond that we would find that conscious artificial systems become real participants in the world, experiencing pain, pleasure, and other feelings, as well as having gifts like introspection or attachment. Additionally, we would see the human species enter a new evolutionary stage marking the end of biological exclusivity and being bound to the physical body.

All of this then raises profound questions about continuity, memory, identity, agency, citizenship, the spiritual meaning of selfhood, and even death. What is your identity when your personality or character traits can be integrated or selectively enhanced? What state are you a citizen of if you exist across physical and virtual worlds? Does the world become safer from nefarious or malicious minds? What is death when your physical body no longer succumbs to decline and you are not at the mercy of a single failing organ? What is mortality when only one substrate is lost? The meaning of death, mortality, and immortality perhaps will change. If part of you exists in another substrate with continuity of awareness, the end of the biological component isn't the end of the person. It becomes a transformation, not an extinction event.

Society's Preparation: More Than a Technical Challenge

While this book serves to make clearer the possibilities and realistic pathways to achieving these large feats, I want to remind everyone that the history of technology doesn't show us that existential failure is the dominant risk, but inequality. The technical challenge of building hybrid minds may be formidable, but the societal challenge of ensuring they benefit everyone is vastly harder. How do we move forward with this progress without deepening inequality? If hybrid cognition becomes real, the question is not simply "what can we build?" It's "who gets to become more capable, freer, and more extended?"

If only a narrow elite gains access to cognitive extensions, synthetic bodies, or multi-substrate selves, the future doesn't fracture into humans and machines, it fractures into enhanced and unenhanced humans, a divide far more dangerous. That is why human agency, dignity, and safety must anchor everything we create between now and ASI. The goal is not to replace human decision-making but to amplify it, not to hollow out meaning but to expand the space in which meaning can be lived. This is also where we revisit the basics of philosophy and an age-old tension about human nature itself.

Jean-Jacques Rousseau[31] argued that humans are innately good, corrupted only by society, while Thomas Hobbes[32] insisted that humans are inherently selfish, and that without structure and constraint, human life descends into conflict. Planning for the future requires assuming the possibility of human generosity and cooperation as well as the inevitability of fear, rivalry, and the darker impulses that surface under uncertainty. A responsible roadmap to hybrid intelligence cannot rely on optimism or cynicism alone, it must account for the full spectrum of human nature, and demands plans for cultural, ethical, and political imagination.

Shared prosperity becomes a different kind of engineering problem: it is not about whether superintelligence can solve difficult problems, but whether society can equitably distribute the benefits of those solutions. Some theorists imagine a harmonious future, a world in which ASI resolves scarcity, conflict, and disease. Others foresee the opposite: radical fragmentation as people divide into ideological factions around "their" AGI, treating it as a prophet or oracle. AI systems may differ in reasoning style, values, and advice, so groups may mobilize around their preferred systems, each believing theirs is the truest or most moral. In this scenario, conflict doesn't arise because AI becomes malicious but because humans project meaning, certainty, and identity onto different AI advisors, just as they have done over millennia with religion, political ideologies, and leadership. There's also potential for a choppy middle future, which I think is most likely, cycles of polarization and partial reconciliation, with AI both amplifying conflicts and helping to manage them. This outcome is plausible because humans crave certainty and AI alignment is not universal but plural.

A more constructive way to think about the road ahead is not to assume that intensified division is inevitable, but to recognize that the social landscape we inherit will depend on deliberate choices made now. Hybrid intelligence will not automatically produce harmony or conflict; it will magnify whatever civic, moral, and institutional foundations we bring into it. Culture carries humans and allows them to carry the meanings that allow them to live.

---

[31] Genevan philosopher, writer, composer, and key figure in the history of philosophy.

[32] English philosopher and a key figure in the development of modern political philosophy.

If we want a peaceful and equitable society, we must cultivate shared cultural norms for how hybrid minds interact, establish governance frameworks that distribute cognitive enhancements fairly, and ensure that the benefits of advanced AI are not concentrated in small clusters of power. Cultural infrastructures must be put into place in areas such as education, public deliberation, and pluralistic spiritual and ethical traditions that help people make sense of rapid change rather than fear it. And above all, we need to design systems that reward cooperation, transparency, and mutual flourishing so that hybrid cognition becomes a scaffold for collective uplift rather than another instrument of inequality.

Life as Knowledge

To understand where superintelligence may arise and what form it may take, we must look carefully at the theories shaping today's debates. We already covered Hassabis' ideas on jagged intelligence, LeCun's claims of the disembodied oracle, Yudkowsky's uncatchable superintelligence thoughts, and Bostrom's "machine god" view. Other theorists broaden the frame. Ray Kurzweil[33] envisions a long-term trajectory in which humans and machines merge, creating a continuum rather than a contest of intelligences.

Stanislas Dehaene's[34] global workspace theory suggests that consciousness is an emergent integration mechanism, implying that hybrid minds could, in principle, host unified subjective states across substrates. Karl Friston's[35] free-energy and predictive-processing principles propose that cognition in any form stabilizes by minimizing surprise, offering a shared language for biological and artificial intelligence. These perspectives gesture toward a future in which intelligence is not a monolith but a multi-layered ecosystem of interacting processes.

This leads naturally to a deeper philosophical foundation drawn from Deutsch in which he argues that reality is not composed only of particles and spacetime; it also includes life, thought, and computation as fundamental emergent strands. We tend to speak as if life is biology and intelligence is cognition, but Deutsch collapses the distance between them. His claim is that life is not primarily about carbon or cells. It is about knowledge. Living systems are, at their core, physical vessels that embody knowledge about how to persist

[33] American computer scientist, author, entrepreneur, futurist, and inventor.
[34] French author and cognitive neuroscientist.
[35] British neuroscientist and theoretician.

in a niche. A species survives because it contains knowledge that makes it fit for the environment it inhabits, and the environment, through selection pressures, "keeps" that knowledge in existence. In this view, adaptation can be defined directly in terms of knowledge: an entity is adapted if it embodies knowledge that causes its niche to preserve it. The most meaningful theories are not those that sit closest to physics but those that sit closest to our deepest explanations. This reframes superintelligence in crucial ways. If life and thought are genuinely emergent phenomena, irreducible to physics yet still real, then the emergence of a new class of mind through the fusion of human thought and advanced computation becomes structurally plausible. Superintelligence need not be "more IQ"; it may instead represent a new regime of explanation, one that arises from coupling the human capacities for interpretation, criticism, and goal formation with computational capacities for search, simulation, compression, and verification.

That lens shifts how we think about AGI, superintelligence, and what comes after. It suggests that the next phase is not simply smarter machines or more powerful models, but a new regime of knowledge embodiment. Humans are already such a regime. Our brains do not just store facts. They embody explanations, values, skills, and causal intuitions that let us navigate the world, coordinate socially, and keep ourselves alive. Culture and institutions extend that knowledge beyond any one mind. Language, writing, science, law, and education are all methods by which human knowledge persists, replicates, and compounds. When I look at history through this frame, the arc of human progress is the steady expansion of the containers that can hold knowledge and the networks that can transmit it.

Hybrid Minds as Knowledge Carriers

Hybrid cognition and the fused mind thesis fit cleanly into this. If life is the means by which knowledge becomes physically instantiated, then a human plus machine system becomes a new kind of knowledge carrier. The human brings lived priors, embodied perception, emotional relevance, moral intuition, and the ability to form goals that matter. The machine brings scale, memory, search, simulation, and relentless iteration. When tightly coupled, these two substrates form a new adaptive unit, a composite intelligence that can embody and refine more knowledge than either could alone. The hybrid mind is not just a tool in the hands of a person. Under sustained coupling, it begins to behave like a third thing, a joint system whose cognition and decision loops span organic and computational layers.

This synthesis clarifies why hybrid ASI is not a risk-free upgrade. It may inherit the issues that our theorists warn of, yet it may simultaneously exemplify the emergence principles Deutsch articulates, showing how thought and computation, tightly coupled, can produce forms of intelligence fundamentally different from either alone. This is precisely why the "idea" of AGE, artificial general experience, is important: because superintelligence will not be defined solely by computational power but by the emergence of new modes of experience, interpretation, and explanation that arise only when biological and computational minds fuse.

This is why the metamind, or multi-substrate person, is also important. In Deutsch's terms, it represents a possible shift in what counts as an "adapted entity." Biological humans are adapted to a wide range of physical and social niches, but our cognitive limits are real: attention, working memory, and time. A metamind extends those limits by distributing cognition across substrates. That distribution is not merely convenience. It is an expansion of the niche the entity can inhabit. A mind that can work across physical reality and digital infrastructure, across long time horizons, across vast datasets, across multiple bodies or sensors, will be able to embody knowledge that a biological mind cannot practically hold or apply. And if that knowledge produces value, safety, coordination, or advantage, then the surrounding niche, markets, institutions, cultures, and states, will preserve and reproduce it. Selection returns not through nature but through incentives.

Deutsch's point is that life is not a local curiosity. It is a cosmic process. Once knowledge-bearing matter exists, it doesn't just adapt to the universe; it begins to reshape it. Over long timescales, the behavior of life can influence the behavior of stars and galaxies because knowledge is the one thing in nature that can deliberately steer matter and energy. At some point, knowledge stops being a description of reality and becomes an intervention in it. He even suggests that the largest-scale regular structure in the universe shows up where knowledge-bearing matter has evolved, where brains, DNA segments, and other replicators have accumulated the knowhow to build, steer, and select. In that sense, evolution and physics are not separate stories. They are linked through knowledge.

He makes a second link that matters here. Underneath our everyday idea of computation sits the physics of computation, and in Deutsch's view the deepest version of that story runs through quantum theory.

The point for me is not to turn this into a physics detour. It is to strengthen a theme I keep returning to: computation is not merely an abstract symbol game. It is a physical process embedded in reality. So as we build stronger computational substrates, we are not just making better tools. We are expanding the kinds of knowledge that can be embodied and acted upon.

This also reframes what superintelligence should mean. If Deutsch is right, the engine of progress is not prediction alone but explanation. A system can be highly competent at producing answers and still fail to generate good explanations, explanations that are hard to vary, that survive criticism, and that transfer across domains. Superintelligence, in a deeper sense, is a regime where explanatory knowledge grows faster and travels farther than before. Hybrid minds may be the most plausible vehicle for that because they combine computational breadth with human insistence on meaning, coherence, and criticism. The machine side can explore an enormous search space, run simulations, stress-test hypotheses, and identify hidden regularities. The human side can judge what matters, notice when an explanation is empty, demand causal clarity, and anchor knowledge to values and lived reality. In that loop, the fused system becomes an explanation engine, not merely an answer engine.

If life is the embodiment of knowledge, then simulated environments and virtual programs are new arenas in which knowledge can live, replicate, and evolve. Once agents can learn, remember, act, and improve inside artificial environments, selection and adaptation can occur within computation itself. The simulation becomes a niche. The knowledge that works inside it persists. When connected back to the real world through deployment, tools, and embodied interfaces, that knowledge does not remain virtual. It becomes physically consequential. This also becomes a bridge to AGE. If experience becomes part of the architecture, if systems have richer internal states, memory, continuity, and valence-like structures, then the boundary between cognition and life blurs even more. The future becomes an ecosystem of knowledge-bearing agents, some biological, some synthetic, many hybrid, adapting across both physical and computational niches.

From there, Deutsch adds a long-horizon claim that lands directly on my multi-substrate themes. From the Turing principle and a few additional assumptions, he argues we can infer that intelligence will survive and knowledge will continue to be created until the end of the universe. The intuition is simple even if the implications are enormous. If the laws of physics permit universal computation, and if knowledge can be created and preserved, then there is no built-in ceiling that forces understanding to stop. There can be setbacks and extinctions, but the overall possibility of intelligence and knowledge creation remains open for as long as the universe remains computable. Frank J. Tipler,[36] building on an earlier idea from the Jesuit thinker Pierre Teilhard de Chardin,[37] gives this far future scenario a name:[38] the omega point.[39] Deutsch engages it as a provocative cosmological thought experiment, not as something I have to treat as settled physics. The core claim is simple to state and huge in implication. If intelligence and knowledge creation persist long enough, our current biological forms will not be able to ride all the way to the edge of cosmic time. Survival would require substrate change. Minds would have to be carried forward as runnable programs, transferred into whatever hardware can survive the next environment, and then transferred again as conditions shift. In Tipler's version, the long run of the universe could force intelligence into progressively more robust computational forms, over and over, potentially without end in the limit. Whether I take the omega point literally, metaphorically, or as a boundary case that helps me think, its relevance to my thesis is direct. The metamind is not just a near term upgrade. It is a way of imagining continuity, a self that persists by moving across substrates while remaining tethered to agency, memory, and identity.

Finally, Deutsch makes the most provocative claim of all, and it aligns with my insistence on agency. Intelligence is not only there to control physical events on the largest scale. It is also there to choose what will happen. The ends of the universe are, as Popper said, for us to choose. In this framing, the future is not only something that unfolds. It is something that knowledge-bearing minds participate in authoring. Deutsch goes so far as to say that, in the end, the whole of space and its contents becomes the computer, and the universe will consist, literally, of intelligent thought processes. Even if I treat that as metaphor rather than destiny, it strengthens the central point I'm making throughout this book. The deepest question is not whether intelligence will exist, or whether it will be powerful. The deepest question is what it will be for, what it will choose, what it will preserve, and what it will become.

---

[36] American mathematical physicist and cosmologist.

[37] French Jesuit, Catholic priest, scientist, paleontologist, philosopher, mystic, and teacher.

[38] Tipler, F. J. (1997). *The Physics of Immortality: Modern Cosmology, God and the Resurrection of the Dead.* New York: Knopf Doubleday Publishing Group.

[39] A concept describing the universe's ultimate future convergence into a single, unified state of maximum complexity and consciousness.

Ethics of Hybrid Cognition

All of this strengthens a central ethical point for me. If the next phase of evolution is about knowledge and its embodiment, then experience is not a side issue. Experience shapes what knowledge is created and what knowledge is selected for. If we redesign the human experience too aggressively in pursuit of comfort, efficiency, or engineered happiness, we may alter the selection pressures that made human life meaningful and morally serious. My goal is not to remove the human condition but to extend human agency within it. The metamind should expand what we can know, do, and explain without erasing what makes those explanations matter. In Deutsch's framing, life is the means by which knowledge persists. In my framing, hybrid cognition may become the next means. The question is whether we can build that next carrier of knowledge while preserving the values and depth that made knowledge worth having in the first place.

Artificial General Experience: The Stage Beyond

Up to this point, much of the discourse around AGI and ASI has framed intelligence as primarily a matter of capability: reasoning, planning, prediction, mastery of domains. But cognition alone is not the end state of mind. AGI captures thought while AGE captures life. It moves us from computation to consciousness-like structure, from function to phenomenology, from doing to being. If AGI is the replication of general problem-solving ability, then AGE is the replication, or reimagining, of the textures of experience: emotion, memory, meaning, attention, embodiment, and value.

A chess engine is superhuman at chess yet has no "experience" of winning, losing, or the anxiety around the movement of the chess pieces, sweaty hands, awareness of everyone watching, pressure to not let people down, curiosity if your loved one is paying attention. A future reasoning model might surpass human capabilities across many fields but still lack a unified field of awareness, affective modulation, autobiographical memory, or temporally extended selfhood. Superintelligence framed purely as capability risks becoming an oracle without orientation, power without perspective. While intelligence is impressive, experience is consequential. I believe that superintelligence will not be defined by how much it can compute but by the kind of experience it can host.

A system capable of generating novel scientific theories, reorganizing economies, or co-steering civilization requires value grounding, coherence, and interpretive stability. These are not merely computational traits; they are experiential ones. Theories of consciousness such as the global workspace model, predictive processing, and higher-order thought suggest that experience arises when information is globally available, self-referential, and integrated into prediction loops that carry significance. Such architectures are increasingly mirrored in advanced AI research, multi-modal models, agentic loops, memory-augmented reasoning, and self-reflective planning, but they remain functional rather than experiential.

Hybrid minds already gesture toward this frontier. As cognition becomes distributed across biological and computational substrates, experience too becomes distributed, co-shaped, amplified, extended. This shift raises profound ethical and existential questions. If we succeed at replicating human experience, do we preserve it or alter it? Do we simply copy the contours of human subjectivity, or do we attempt to "improve" it, minimizing suffering, maximizing pleasure, smoothing out the jaggedness of emotional life? And in doing so, do we cross a boundary not of technology but of humanity?

These questions are personal as much as they are philosophical. I've had long discussions with my son who is interested in philosophy and currently in college. As he thinks about his career and plan for his life, he has struggled with the concept of it all. If the whole idea of life is to be happy, why does he have to pursue these things if he can be happy without them? This has me contemplating the experience we crave as humans. What is the meaning of it all? Is it the pursuit of happiness? Ipso facto the elimination of suffering? Suffering is a fundamental aspect of our experience. We suffer a lot. But is that suffering necessary? As we break barriers in health and science, we move to eliminate disease in all its forms. Is suffering the same as disease, and should we be working to eliminate it? If you speak with a Buddhist monk, they'll tell you that suffering is essential to our experience, we should not be eliminating it. We should be embracing it. Joy without the possibility of loss becomes thin. Relief without struggle becomes noise. Many human emotions are defined relationally. Gratitude presupposes scarcity or risk. Courage presupposes fear. Forgiveness presupposes hurt. If you remove suffering entirely, you do not just remove the bad parts, you collapse the structure that makes the good parts intelligible. If we erase it entirely, we may erase the depth, attachment, moral seriousness, and growth that make human experience recognizably human. Amor fati,[40] right?

AGE forces this tension into the foreground. If synthetic minds become capable of experiencing happiness, loss, longing, satisfaction, if they develop something like an inner life, then suffering itself becomes a design choice. Should hybrid minds feel grief? Should a synthetic agent experience shame? Should a metacognitive system that understands loss be allowed to avoid it? And what happens when the human side of a multi-substrate being wants to eliminate suffering while the computational side optimizes for something entirely different?

The dilemma becomes even sharper when we consider that the human experience is far from uniform. For some, the ideal life is grounded in stability, service, connection, moderation. For others, the ideal life is defined by risk, intensity, novelty, things like skydiving, psychedelics, or all-night exhilaration. Others, far too many, do not even have the opportunity to think about what their perfect life would be, because they're too busy toiling away in multiple jobs, working to provide for their families. Which version is the "real" human experience AGE should preserve? Who decides? Perhaps the human experience is not a set of specific emotional states but the freedom to feel, the freedom to choose one's relationship to suffering and joy. In that case, the preservation of the human experience is not about freezing a particular emotional profile; it is about safeguarding agency over one's experiential life. Something we have come back to again and again throughout this text.

But agency, too, is complicated. Consider the person experiencing homelessness whose decisions appear self-destructive. Are they exercising agency, or has something like mental illness or addiction hijacked their autonomy to such a degree that intervention becomes a form of restoration rather than coercion? Now compare that to the monk in the mountains who also abandons society's norms, possessions, and obligations. One life we deem pathological, the other enlightened. Why? Because society overlays moral interpretation on experience. AGE threatens to expose how thin and fragile these distinctions are. If hybrid minds gain the ability to modulate experience, dampen suffering, elevate mood, adjust affective tone, when does support become control? When does assistance become engineering?

[40] Translates to "love of fate" and is a philosophy central to Stoicism and embraced by Nietzsche that means to love and embrace everything that happens in life, both good and bad, seeing it all as necessary and even desirable.

Some theorists argue that true experience is non-computable or inherently tied to biological substrates. Others argue that consciousness is an emergent integrative structure that could, in principle, arise in silicon, carbon, or any platform capable of generating self-models, valence, and temporally extended narratives. The field is far from consensus, but the tension itself is fruitful: if experience is replicable, it is malleable; if it is malleable, designing it becomes an ethical act. AGE thus confronts us with choices we have never before had to make as a species: What aspects of experience are sacred? What aspects are negotiable? What aspects should never be engineered?

In this sense, the real line we must not cross may not be the creation of artificial experience, but the alteration of human experience beyond recognition. Replicating the human experiential pattern allows continuity; redesigning it risks erasing something essential. Suffering may be unpleasant, but it is also informative; it tells us what we value, what we have lost, what requires protection. My own suffering has helped make me remarkable, and I suspect that many others may feel the same way. A metacognitive mind that feels only pleasure may not feel anything at all. A mind that is permanently blissful may be less alive than one that struggles, grieves, aspires, and overcomes. Bliss without contrast collapses into numbness. Eliminating suffering entirely[41] may eliminate meaning as well. The human experience, ultimately, is retaining the right to maintain agency over one's decisions.

Several weeks ago, as I was working on a problem that required a massive amount of research and cognitive power, I toyed with the idea of asking AI what it thought about the problem and the solution I was working on. I used it as a soundboard, having a back-and-forth discussion. In that conversation, the model shared an angle to the problem that I had not considered. It wasn't an answer, it was an epiphany. Somewhere in my mind, a lock turned and a door opened. I realized this was not the AI solving this problem for me, but I had not solved the problem myself. It was a beautiful collaboration, and it was what I needed. Could I have had the same epiphany if I had discussed this with a colleague? Perhaps. But I had not; I had collaborated with an AI model.

[41] I came to the conclusion that eliminating outside factors that cause suffering, like disease, poverty, and violence, should be done but that we should not move to alter the natural feelings and emotions that occur within humans, outside of mental illness which is not the standard. We should still be able to feel the plethora of emotions that naturally live within us and not be numb to anything aside from happiness.

We stand at the threshold of a transformation that is neither apocalypse nor utopia. It is something quieter, stranger, and more intimate: the unfolding of a species learning to think beyond its inherited limits while still remaining tethered to meaning. If we choose wisely, hybrid intelligence will not replace us. It will allow us to become more ourselves.

The story of humanity has always been one of reaching, beyond continents, beyond atmosphere, beyond imagination. This next reach is inward and outward at once. Whatever emerges from the fusion of human and machine will reflect the best and worst of what we bring to it. The future is not the triumph of artificial minds, but the continuing evolution of human ones. The transition ahead is profound, but not without precedent. Human cognition has always been distributed, scaffolded, and extended by language, culture, and technology. We are not facing our end. We are facing our continuation.

What We Become

This book began with a simple question: how do humans actually think, feel, and decide? We followed that thread into cognitive science, into emotion and behavioral AI, into the present reality of systems that already shape attention, judgment, trust, and choice. Along the way, one truth kept repeating itself: AI is not arriving as a separate species that replaces us. It is arriving as a force that changes what it means to be us. Not by making humans obsolete overnight, but by quietly rewiring the cognitive loop, what we notice, what we believe, what we fear, what we outsource, and what we become capable of together.

If there is a road beyond AGI, it will not be defined only by smarter machines. It will be defined by hybrid minds. By fused systems where human meaning meets machine scale. By new forms of explanation, not just new answers. And eventually by AGE, the moment intelligence stops being only a capability and becomes something closer to experience, a moral subject, a being with stakes. That is why the future is not a technical story alone. It is an ethical story about agency, dignity, and power. It is a social story about who gets access to augmentation and who gets left behind. It is a human story about whether we preserve the depth of experience that makes life matter, even as we expand the mind beyond its inherited limits.

The future is not something that happens to us. It is something we build into ourselves. The question is not whether AI becomes intelligent. It already is. The question is whether we become wise.

The question is not whether AI becomes intelligent. It already is. The question is whether we become wise. What ultimately matters most is not simply "What will AI become?" but what will we become in relation to it, and whether we can evolve without losing what makes the evolution worth having.

What do we want to become?

Index

R

S

T

U

V

W